Studies in Computational Intelligence

Volume 656

Series editor

Janusz Kacprzyk, Polish Academy of Sciences, Warsaw, Poland
e-mail: kacprzyk@ibspan.waw.pl

About this Series

The series "Studies in Computational Intelligence" (SCI) publishes new developments and advances in the various areas of computational intelligence—quickly and with a high quality. The intent is to cover the theory, applications, and design methods of computational intelligence, as embedded in the fields of engineering, computer science, physics and life sciences, as well as the methodologies behind them. The series contains monographs, lecture notes and edited volumes in computational intelligence spanning the areas of neural networks, connectionist systems, genetic algorithms, evolutionary computation, artificial intelligence, cellular automata, self-organizing systems, soft computing, fuzzy systems, and hybrid intelligent systems. Of particular value to both the contributors and the readership are the short publication timeframe and the worldwide distribution, which enable both wide and rapid dissemination of research output.

More information about this series at http://www.springer.com/series/7092

Roger Lee
Editor

Computer and Information Science

Editor
Roger Lee
Software Engineering and Information
Technology Institute
Central Michigan University
Mount Pleasant
USA

ISSN 1860-949X ISSN 1860-9503 (electronic)
Studies in Computational Intelligence
ISBN 978-3-319-82045-3 ISBN 978-3-319-40171-3 (eBook)
DOI 10.1007/978-3-319-40171-3

Softcover reprint of the hardcover 1st edition 2016

Printed on acid-free paper

This Springer imprint is published by Springer Nature
The registered company is Springer International Publishing AG Switzerland

Foreword

The purpose of the 15th IEEE/ACIS International Conference on Computer and Information Science (ICIS 2016) held on June 26–29, 2016 in Okayama, Japan, was to bring together researchers, scientists, engineers, industry practitioners, and students to discuss, encourage, and exchange new ideas, research results, and experiences on all aspects of Applied Computers and Information Technology, and to discuss the practical challenges encountered along the way and the solutions adopted to solve them. The conference organizers have selected the best 12 papers from those papers accepted for presentation at the conference in order to publish them in this volume. The papers were chosen based on review scores submitted by the members of the program committee and underwent further rigorous rounds of review.

In Chap. "The Drill-Locate-Drill (DLD) Algorithm for Automated Medical Diagnostic Reasoning: Implementation and Evaluation in Psychiatry", D.A. Irosh P. Fernando and Frans A. Henskens propose a drill-locate-drill (DLD) algorithm models the expert clinician's top-down diagnostic reasoning process, which generates a set of diagnostic hypotheses using a set of screening symptoms, and then tests them by eliciting specific clinical information for each differential diagnosis.

In Chap. "Implementation of Artificial Neural Network and Multilevel of Discrete Wavelet Transform for Voice Recognition", Bandhit Suksiri and Masahiro Fukumoto present an implementation of simple artificial neural network model and multilevel of discrete wavelet transform as feature extractions. The proposed method offers a potential alternative to intelligence voice recognition system in speech analysis-synthesis and recognition applications.

In Chap. "Parallel Dictionary Learning for Multimodal Voice Conversion Using Matrix Factorization", Ryo Aihara, Kenta Masaka, Tetsuya Takiguchi, and Yasuo Ariki propose a parallel dictionary learning for multimodal voice conversion. Experimental results showed that their proposed method effectively converted multimodal features in a clean environment.

In Chap. "Unanticipated Context Awareness for Software Configuration Access Using the getenv API", Markus Raab introduces an integrated unmodified applications into a coherent and context-aware system by instrumenting the getenv API. The results show that getenv is used extensively for variability. The tool has acceptable overhead and improves context awareness of many applications.

In Chap. "Stripes-Based Object Matching", Oliver Tiebe, Cong Yang, Muhammad Hassan Khan, Marcin Grzegorzek, and Dominik Scarpin propose a novel and fast 3D object matching framework that is able to fully utilize the geometry of objects without any object reconstruction process. They show that the proposed method achieves promising results on some challenging real-life objects.

In Chap. "User Engagement Analytics Based on Web Contents", Phoom Chokrasamesiri and Twittie Senivongse propose a set of web content-based user engagement metrics that are adapted from existing web page-based engagement metrics. In addition, the proposed metrics are accompanied by an analytics tool which the web developers can install on their websites to acquire deeper user engagement information.

In Chap. "Business Process Verification and Restructuring LTL Formula Based on Machine Learning Approach", Hiroki Horita, Hideaki Hirayama, Takeo Hayase, Yasuyuki Tahara, and Akihiko Ohsuga use LTL verification and prediction based on decision tree learning for verification of specific properties. Furthermore, they help writing properly LTL formula for representing the correct desirable property using decision tree construction. They conducted a case study for evaluations.

In Chap. "Development of a LMS with Dynamic Support Functions for Active Learning", Isao Kikukawa, Chise Aritomi, and Youzou Miyadera propose a new type of learning management system (LMS), which gives teachers more freedom in planning active learning. They named it "Dynamic LMS (DLMS)". These linked functions enable teachers to plan a variety of activities and, consequently, create more effective active learning in their classes.

In Chap. "Content-Based Microscopic Image Retrieval of Environmental Microorganisms Using Multiple Colour Channels Fusion", Yanling Zou, Chen Li, Kimiaki Shiriham, Florian Schmidt, Tao Jiang, and Marcin Grzegorzek developed an EM search system based on content-based image retrieval (CBIR) method by using multiple colour channels fusion. The system searches over a database to find EM images that are relevant to the query EM image.

In Chap. "Enhancing Spatial Data Warehouse Exploitation: A SOLAP Recommendation Approach", Saida Aissi, Moha med Salah Gouider, Tarek Sboui, and Lamjed Ben Said present a recommendation approach that proposes personalized queries to SOLAP users in order to enhance the exploitation of spatial data warehouses. The approach allows implicit extraction of the preferences and needs of SOLAP users using a spatial-semantic similarity measure between queries of different users.

In Chap. "On the Prevalence of Function Side Effects in General Purpose Open Source Software Systems", Saleh M. Alnaeli, Amanda Ali Taha, and Tyler Timm examine the prevalence and distribution of function side effects in general-purpose software systems. An analysis of the historical data over a 7-year period for 10 systems shows that there are a relatively large percentage of affected functions over the lifetime of the systems.

In Chap. "aIME: A New Input Method Based on Chinese Characters Algebra", Antoine Bossard describes a novel way to approach Chinese characters: the algebraic. They propose a concrete implementation of an input method editor (IME). From the results obtained, it is clear that the proposed input method brings significant improvement over conventional approaches.

It is our sincere hope that this volume provides stimulation and inspiration, and that it will be used as a foundation for works to come.

June 2016

Masahide Nakamura
Kobe University, Japan

Contents

Contributors

Ryo Aihara Graduate School of System Informatics, Kobe University, Kobe, Japan

Saida Aissi SOIE Laboratory, High Institute of Management, Tunis, Tunisia

Saleh M. Alnaeli Department of Computer Science, University of Wisconsin-Fox Valley, Menasha, WI, USA

Yasuo Ariki Graduate School of System Informatics, Kobe University, Kobe, Japan

Chise Aritomi Tokoha University, Fuji-shi, Shizuoka, Japan

Lamjed Ben Said SOIE Laboratory, High Institute of Management, Tunis, Tunisia

Antoine Bossard Graduate School of Science, Kanagawa University, Hiratsuka, Kanagawa, Japan

Phoom Chokrasamesiri Faculty of Engineering, Department of Computer Engineering, Chulalongkorn University, Bangkok, Thailand

D.A. Irosh P. Fernando Distributed Computing Research Group, School of Electrical Engineering and Computer Science, School of Medicine and Public Health, University of Newcastle, Callaghan, NSW, Australia

Masahiro Fukumoto School of Information, Kochi University of Technology (KUT), Kami City, Kochi, Japan

Mohamed Salah Gouider SOIE Laboratory, High Institute of Management, Tunis, Tunisia

Marcin Grzegorzek Institute for Vision and Graphics, University of Siegen, Siegen, Germany

Takeo Hayase Toshiba Corporation, Tokyo, Japan

Frans A. Henskens Distributed Computing Research Group, Health Behaviour Research Group, School of Electrical Engineering and Computer Science, University of Newcastle, Callaghan, NSW, Australia

Hideaki Hirayama The University of Electro-Communications, Chofu, Tokyo, Japan

Hiroki Horita The University of Electro-Communications, Chofu, Tokyo, Japan

Tao Jiang Chengdu University of Information Technology, Chengdu, China

Muhammad Hassan Khan Institute for Vision and Graphics, University of Siegen, Siegen, Germany

Isao Kikukawa Tokoha University, Fuji-shi, Shizuoka, Japan

Chen Li Institute for Vision and Graphics, University of Siegen, Siegen, Germany

Kenta Masaka Graduate School of System Informatics, Kobe University, Kobe, Japan

Youzou Miyadera Tokyo Gakugei University, Koganei-shi, Tokyo, Japan

Akihiko Ohsuga The University of Electro-Communications, Chofu, Tokyo, Japan

Markus Raab Institute of Computer Languages, Vienna University of Technology, Vienna, Austria

Tarek Sboui ESSPCR (UR 11ES15) & CONTOS2, Faculty of Sciences, Department of Geology, Tunis, Tunisia

Dominik Scarpin Institute of Automatic Control Engineering, University of Siegen, Siegen, Germany

Florian Schmidt Institute for Vision and Graphics, University of Siegen, Siegen, Germany

Twittie Senivongse Faculty of Engineering, Department of Computer Engineering, Chulalongkorn University, Bangkok, Thailand

Kimiaki Shiriham Institute for Vision and Graphics, University of Siegen, Siegen, Germany

Bandhit Suksiri Graduate School of Engineering, Kochi University of Technology (KUT), Kami City, Kochi, Japan

Amanda Ali Taha Department of Computer Science, University of Wisconsin-Fox Valley, Menasha, WI, USA

Yasuyuki Tahara The University of Electro-Communications, Chofu, Tokyo, Japan

Tetsuya Takiguchi Graduate School of System Informatics, Kobe University, Kobe, Japan

Oliver Tiebe Institute for Vision and Graphics, University of Siegen, Siegen, Germany

Tyler Timm Department of Computer Science, University of Wisconsin-Fox Valley, Menasha, WI, USA

Cong Yang Institute for Vision and Graphics, University of Siegen, Siegen, Germany

Yanling Zou Chengdu University of Information Technology, Chengdu, China

The Drill-Locate-Drill (DLD) Algorithm for Automated Medical Diagnostic Reasoning: Implementation and Evaluation in Psychiatry

D.A. Irosh P. Fernando and Frans A. Henskens

Abstract The drill-locate-drill (DLD) algorithm models the expert clinician's top-down diagnostic reasoning process, which generates a set of diagnostic hypotheses using a set of screening symptoms, and then tests them by eliciting specific clinical information for each differential diagnosis. The algorithm arrives at final diagnoses by matching the elicited clinical features with what is expected in each differential diagnosis using an efficient technique known as the orthogonal vector projection method. The DLD algorithm is compared with its rival select-test (ST) algorithm and its design/implementation in psychiatry, and evaluation using actual patient data is discussed.

Keywords Drill-locate-drill (DLD) algorithm · Orthogonal vector projection method · Medical expert systems · Automated diagnostic reasoning

1 Introduction

There have been various approaches to automate medical diagnostic reasoning including theoretical models and their implementations as medical expert systems. Some of the theoretical models include Certainty Factor model [1], Parsimonious Covering Theory [2], Process Model for diagnostic reasoning [3], Information Processing Approach [4], scheme-inductive reasoning [5], backward and forward

D.A.I.P. Fernando (✉)
Distributed Computing Research Group, School of Electrical Engineering & Computer Science, School of Medicine and Public Health, University of Newcastle, Callaghan, NSW, Australia
e-mail: irosh.fernando@uon.edu.au

F.A. Henskens
Distributed Computing Research Group, Health Behaviour Research Group, School of Electrical Engineering & Computer Science, University of Newcastle, Callaghan, NSW, Australia
e-mail: frans.henskens@newcastle.edu.au

R. Lee (ed.), *Computer and Information Science*,
Studies in Computational Intelligence 656, DOI 10.1007/978-3-319-40171-3_1

reasoning [6], pattern recognition [7], and hypothetico-deductive reasoning [8]. There have also been approaches that are based on probability theory [9, 10, 11] and Fuzzy logic [12, 13, 14]. Nonetheless, some of the major projects such as INTERNIST-1 and CADUCEUS, which were aimed at automating medical diagnostic reasoning based on some of these theories, resulted in failure after a decade long development effort [15]. Whilst there are many factors including lack of consideration of organisational structure in which the system is used [16] and difficulties in getting clinicians to use the system [17] that can be considered as contributory factors, lack of a proper theoretical foundation and effective computational techniques can also be considered as a major reason for such failures.

Medical diagnostic reasoning consists of two stages (1) an informed search for clinical information driven by diagnostic hypothesis in order to include and exclude various diagnostic possibilities; (2) arriving at differential diagnoses with the likelihood of each diagnosis based on clinical information that was gathered through the search process. Whilst most of the above stated approaches lack an exhaustive search mechanism, the existing computational approaches, which include rule based systems and probabilistic approaches based on Bayes theorem for deriving at diagnostic conclusions, are not effective. The process of arriving at diagnostic conclusions involves a complex process of matching the elicited clinical information with various diagnostic criteria for differential diagnosis. The large number of rules and the large number of joint probability distributions that are required to represent medical knowledge can become unmanageable, and additionally rule based-systems are not effective in handling missing information.

Of all the proposed theories of medical diagnostic reasoning, the Select and Test (ST) model [18] can be considered as the most complete model that captures the diagnostic reasoning process of expert clinicians. The ST model uses the logical inferences: abduction, deduction, induction, and abstraction [19]. It is worth mentioning that the ST model, along with the logical inferences, has been used as the theoretical foundation for teaching clinicians diagnostic reasoning in psychiatry. As an effective computational techniques for matching elicited clinical information with standard diagnoses and deriving diagnostic conclusions, a method known as orthogonal vector projection method [20] has been introduced. Additionally, a diagnostic reasoning algorithm known as the ST algorithm, which is based on the ST model and augmented with the orthogonal vector projection method [21], has been demonstrated in diagnostic reasoning in psychiatry [22], which is a subdomain of clinical medicine. The ST algorithm is able to perform an exhaustive search by iterating the inferences abduction, deduction, and abstraction, thus resulting in a computational complexity of $O(n^3)$ [22].

This paper introduces and evaluates Drill-Locate-Drill (DLD), a rival algorithm that uses the same logical inferences in a strictly top-down manner without iterating them. As a result, DLD has an improved efficiency of $O(n^2)$ at a risk of missing diagnoses since it is unable to perform an exhaustive search. However, this risk can be minimised as described in the following section.

2 Modelling Inferences in Medical Diagnostic Reasoning and the Top-Down Reasoning Algorithm

Based on the clinical expertise of the first author (who is a practising psychiatrist) and his observations on how psychiatrists perform diagnostic reasoning, top-down reasoning is the approach most often used in clinical psychiatry. Such a top-down approach is further suggested and facilitated given that psychiatric diagnoses are classified in a hierarchical manner in the Diagnostic and Statistical Manual of Mental Disorders (DSM)-V [23] and the International Classification of Diagnoses (ICD)-10 [24], which are the 2 main standard diagnostic classification systems in psychiatry.

The DLD algorithm models this top-down diagnostic reasoning process using the above stated inferences, about which the reader may refer to [19, 25] for further detail. Whilst the knowledge structure that is used by the DLD algorithm described in this paper is conceptualised as a 3-layer hierarchical graph consisting of sets $ScreeningSymptoms = \{x_1, x_2, \ldots, x_k\}$, diagnoses $Diagnoses = \{d_1, d_2, \ldots, d_m\}$, and more specific clinical features $ClinicalFeatures = \{x_{k+1}, x_{k+2}, \ldots, x_n\}$ where k, m, n are element of natural numbers N, the general DLD algorithm uses multi-layers including a layer of diagnostic categories and a layer of clinical feature attributes [26]. It is thus consistent with hierarchical conceptualisation of clinical psychiatry knowledge [25]. The directed edges between the screening symptoms and diagnoses represent the posterior probability $(d_i \mid x_j)$, which is the probability of diagnosis d_i given the (secerning) symptom x_j where $i = 1, 2, \ldots, m$ and $j = 1, 2, \ldots, k, k+1, \ldots, n$.

The first step of diagnostic process starts with the algorithm searching for the presence of screening symptoms in the patient. The process of mapping clinical features that are communicated by the patient and observed by the clinician (e.g. emotional state observed via facial expressions, physical activity level, and speech) into items in *ClinicalFeatures* is a complex task known as abstraction [19]. Also, it requires quantification of each clinical feature found in the patient in terms of its severity. A realistic modelling and implementation of abstraction will involve complex human computer interactions using natural language and multimodal sensory processing, and is beyond the scope of this work. Therefore, for the purpose of the DLD algorithm, at a more abstract level all clinical features of the patient including those that are not yet known to the clinician and need to be sought, and their severities, are represented in two sets *SymptomProfile* and *SeverityProfile* respectively. The process of gauging the severity of each symptom can be conceptualised using the function $q\colon ClinicalFeatures \rightarrow R$, where $q(x)$ represent the severity of clinical feature $x \in ClinicalFeatures$ The abstraction will involve checking if any $x \in ScreeningSymptoms$ is a member of *PatientProfile* and finding the corresponding $q(x) \in SeverityProfile$. For the screened clinical features x that were found, x is stored in a set *SymptomsFound* whilst the corresponding severity $q(x)$ is stored in a set *SeveritiesFound*.

The second step of the algorithm uses $P(d_i \mid x_j)$ to derive likely diagnoses (i.e. differential diagnoses) based on screening symptoms found in the patient via abduction where $x_i \in SymptomsFound$ It is important to note that even though some diagnoses are rare (i.e. having a low $P(d_i \mid x_j)$), they can be more critical than some of the common diagnoses, and therefore should still be considered in differential diagnosis. This is modelled by assigning each diagnosis a real value according to its relative criticality using the function $c\colon Diagnoses \rightarrow R$, where $c(d_i)$ is the relative criticality of diagnosis d_i Therefore, the algorithm uses two threshold values t_P and t_C in determining differential diagnoses according to their probability and criticality respectively. The algorithm stores differential diagnoses in the set *DifferentialDiagnoses*.

In establishing a diagnosis each symptom has a relative importance in diagnostic criteria. For example, according to DSM-V diagnostic criteria for Major Depressive Disorder it is essential to have either of depressed mood or loss of interest (or pleasure) as core symptoms with five or more symptoms altogether at least. In order to capture this relative importance of each symptom in relation to a given diagnosis, we have introduce a function $ClinicalFeatures \times Diagnoses \rightarrow [0, 10]$, which assigns each pair of {specific clinical feature and diagnosis} a weight from the interval [0, 10].

Therefore, the third step of the algorithm, which involves deriving the expected symptoms for each differential diagnosis, not only requires the posterior probability $P(x_j \mid d_i)$, which is the probability of having clinical feature x_j given the diagnosis d_i but also $W(x_j, d_i)$, which is the relative importance of x_j in the diagnostic criteria for d_i. Accordingly the edges between the set of specific clinical features and the set of diagnoses represent $w_{ij} = P(x_j \mid d_i) \cdot W(x_j, d_i)$, which is used by the algorithm to derive the expected clinical features for each differential diagnosis via deduction. The expected clinical features that are to be sought in the patient are stored in a set *SymptomsToBeElicited*.

The fourth step is similar to the first step described above, and involves establishing if the expected clinical features for differential diagnoses are found in the patient via abstraction. In the DLD algorithm this simply involves searching the two sets *SymptomProfile* and *SeverityProfile* for each $x \in SymptomsToBeElicited$. If x is found then x is added to *SymptomsFound* and the corresponding $q(x)$ is added to *SeveritiesFound*.

The final step involves matching all the elicited clinical features including their severities with what is expected in each differential diagnosis (i.e. diagnostic criteria) and deriving the likelihood of each diagnosis via induction. The DLD algorithm is augmented with the orthogonal vector projection method [20], which is used to achieve this purpose. Whilst the reader may refer elsewhere for details [20], the DLD algorithm uses the vector $W_i = \, <w_{i1}, w_{i2}, \ldots, w_{in}>$ where w_{ij} is defined above for each diagnosis $i = 1, 2, \ldots, m$ and each clinical feature $j = 1, 2, \ldots, n$, as the standard vector in orthogonal vector projection method. Using the vector $Q = \, <q(x_1), q(x_2), \ldots, q(x_n)>$, which includes elicited and quantified symptoms during the previous steps, the likelihood of each diagnosis d_i denoted as $L(d_i)$ is derived by comparing the similarity of Q to W_i as described in step-5 of the DLD algorithm listed below as Fig. 1. The algorithm handles missing values by setting $q(x_j) = 0$ if x_j is not an element of *SymptomsFound*.

INPUT

Knowledgebase, which is a special bipartite Graph $K_{m,n}$ that has set of vertices $d_1, \ldots, d_m$ as diagnoses and $x_1, x_2 \ldots, x_k$ as screening symptoms and each pair (x_i, d_j) where $i = 1, 2, \ldots, k$ and $j = 1, 2, \ldots, m$ has a directed edge representing $w_{ij} = P(d_j \mid x_i)$; and $x_{k+1}, \ldots, x_n$ as clinical features and each pair (x_i, d_j) where $i = k+1, k+2, \ldots, n$ and $j = 1, 2, \ldots, m$ has a directed edge representing $w_{ij} = P(x_j \mid d_i) \cdot W(x_j, d_i)$.
Profile of clinical features of patient, *SymptomProfile* and *SeverityProfile*
Threshold for probability of diagnoses,t_P
Threshold for criticality of diagnoses, t_C

OUTPUT

The set of likely diagnoses with their likelihood: *DiagnosesIncluded*

3 Implementation

Given that one of the main aims of the implementation was to demonstrate the utility of the algorithm in real world application, it was challenging to design a knowledgebase that is large enough to cover common psychiatric disorders and at the same time small enough to describe in this paper. As a compromise, based on the conceptualisation of clinical knowledge in psychiatry as a hierarchical structure [25], 70 clinical features were defined by clustering the related clinical features as one item. A total number of 44 psychiatric diagnoses were chosen, and the psychiatric diagnoses that can be considered as a union of the chosen diagnoses were excluded. The list of clinical features including screening symptoms and the knowledgebase consisting of w_{ij} is included as a Tables 1 and 2 in the appendix.

The algorithm uses a patient profile with screening symptoms that were found in the patient as input, and then derives differential diagnoses and their expected symptoms as shown in Fig. 2. The quantification each clinical feature was simplified using binary values 0 and 1 for their presence and absence respectively; q: $ClinicalFeatures \rightarrow \{0, 1\}$ (Fig. 3).

4 Evaluation and Results

After obtaining the required ethical approvals from Human Research Ethics Committee of Hunter New England Local Area Health District, a total number of 101 de-identified patient records were collected via three psychiatric teams, each

BEGIN
//initialise empty sets
$SymptomsFound = \emptyset;$
$DifferentialDiagnoses = \emptyset;$
$SymptomsToBeElicited = \emptyset;$

//step-1(abstraction): screening for clinical symptoms, which point to main diagnostic categories
For each $x_j \in ScreeningSymptoms,\ j = 1,2,\ldots,k$ {
If $x_j \in SymptomProfile$,{
add x_j to $SymptomsFound$
add $q(x_j) \in SeverityProfile$ to $SeveritiesFound$.
}
}

//step-2(abduction): derive differential diagnoses for each screening symptom found in patient
For each $x_j \in SymptomsFound$ {
For each d_i where $P(d_i|x_j) \geq t_P\ OR\ C(d_i) \geq t_C,\ i = 1,2,\ldots,m$
Add d_i to $DifferentialDiagnoses$
}

//step-3(deduction): elicit other symptoms that are expected in each differential diagnosis
For each $d_i \in DifferentialDiagnoses$ {
For each x_j where $w_{ij} > t_P$ or $w_{ij} > t_C$ {
If $x_j \notin SymptomsFound$ Add x_j to $SymptomsToBeElicited$.
}
}
// step-4(abstraction): elicit expected symptoms for each differential diagnoses
For each $x \in SymptomsToBeElicited$ {
If $s \in SymptomProfile${
add x to $SymptomsFound$
add $q(x) \in SeverityProfile$ to $SeveritiesFound$.
}
}

//step-5(induction): derive diagnostic conclusion
For each $d_i \in DiagnosesAlreadyElicited$ {

$$L(d_i) = \frac{W_i.Q}{|W_i|^2}$$

Where,

$$W_i.Q = \left(\sum_{j=1}^{n} w_{ij}.q(x_j)\right),$$

$$|W_i|^2 = \sum_{j=1}^{n} w_{ij}^2$$

If $L(d_i) > t_P\ OR\ L(d_i) > t_C$ add d_i to $DiagnosesIncluded$
}
END

Fig. 1 The DLD Algorithm

Table 1 Screening symptoms and clinical features

Index	Screening symptoms
1	Period of depressed mood or psychological distress
2	Period of elevated mood
3	Self-harm/suicidal behaviour attempt
4	Excessive anxiety or severe agitation
5	Disruptive or disorganised behaviour, overactivity or inattention,
6	Anger and aggression
7	Irrational behaviour or thoughts, perceptual abnormalities, disordered speech
8	Cognitive symptoms (e.g. disorientation, impaired consciousness)
9	Somatic symptoms or bodily concerns
10	Binge eating or concerns about body weight
11	Excessive us use of drug and/or alcohol
Index	Clincal features
12	Impairment in social interactions
13	Impairment in communication and language
14	Restricted, repetitive and stereotyped patterns of behaviour, interests, and activities
15	Memory impairment (gradual)
16	Aphasia, apraxia, agnosia, or impairment in executive functioning
17	Presence of systemic conditions that cause cognitive impairment
18	Evidence of cerebrovascular disease
19	Disturbance in consciousness (i.e. disorientation, memory impairment), which fluctuates
20	Regular use alcohol excessively for a prolonged period with evidence of tolerance or withdrawals
21	Presence of physical, social(e.g. work, relationship) or legal problems due to excessive alcohol use
22	Failed attempts to reduce drinking or previous detox/rehab
23	Evidence of alcohol intoxication
24	Regular use THC excessively for a prolonged period with evidence of tolerance or withdrawals
25	Physical, social(e.g. work, relationship), psychological(e.g. psychotic symptoms) or legal problems due to excessive THC use
26	Regular and excessive use of opiates for a prolonged period with evidence of tolerance or withdrawals
27	Presence of physical, social(e.g. work, relationship) or legal problems due to opiate use
28	Attempts to reduce or give up use of opiates (e.g. previous detox/rehab or methadone treatment)
29	Regular and excessive use of amphetamine related substances for a prolonged period with evidence of tolerance or withdrawals
30	Presence of physical, social(e.g. work, relationship) or legal problems due to amphetamine or related substance use
31	Delusions for nearly 1 month or longer

(continued)

Table 1 (continued)

Index	Screening symptoms
32	Hallucinations for nearly 1 month or longer
33	Thought disorder or disorganised behaviour including catatonia for one mo nth or longer
34	Negative symptoms (e.g. flat affect, a motivation) for one month or longer
35	delusions, hallucinations or thought disorder lasting for less than one month
36	Onset of delusions, hallucinations or thought disorder temporarily related to use of THC
37	Onset of delusions, hallucinations or thought disorder temporarily related to use of amphetamine
38	Persistent depressed mood or loss of interest/enjoyment for more than 2 weeks
39	Depressed mood for most days for more than 2 years (not persistent)
40	Loss of appetite or overeating or insomnia or hypersomnia or fatigue
41	Thoughts of guilt or worthlessness or hopelessness or low self-esteem of loss of self-confidence
42	Suicidal thoughts, behaviour or attempts
43	Past history of depression or treatment for depression
44	Elevated or irritable mood or grandiosity
45	Decreased need for sleep or increased activity level or psycho motor agitation
46	Increased speech or racing thoughts
47	Distractibility or impaired concentration
48	Past history of manic episodes or treatment for bipolar affective disorder
49	Recurrent panic attacks with 2 or more symptoms of chest discomfort, nausea, dizziness, feeling dread or loosing control, tingling or hot flushes
50	Anxiety about being in places where panic attacks may occur or avoidance of such places
51	Fear of being humiliated or scrutinised/judged by others and avoidance of social or performance situations
52	Recurrent and persistent obsessions or compulsions
53	Intrusive distressing recollection(e.g. flashbacks, nightmares) of traumatic event or avoidance of its triggers
54	Increased hyperarousal (hypervigilance or exaggerated startle response or impaired concentration or insomnia or irritability)
55	Excessive almost constant worry or apprehensive expectation of worse outcomes about various situations
56	Feeling on edge or muscle tension or restlessness or insomnia
57	Multiple pain, gastrointestinal, sexual and neurological symptoms, which are medically unexplained for several years
58	Medically unexplained unintentionally produced neurological symptoms or deficits (e.g. motor) preceded by a stressor or conflict
59	Preoccupation with fears of having a serious illness for several months
60	Preoccupation with and imagined physical defect or anomaly resulting excessive concern or distress

(continued)

Table 1 (continued)

Index	Screening symptoms
61	Intentional production of physical or psychological symptoms with the motivation to assume a sick role
62	Intense fear of gaining weight with undue influence of body weight on self-evaluation
63	Weight loss resulting physical symptoms or complications (amenorrhoea)
64	Recurrent binge eating with compensatory behaviour (e.g. induced vomiting, fasting, excessive exercise) and undue influence of body weight on self-evaluation
65	Depressed mood or excessive psychological distress or anxiety or insomnia following a major stressful event and relive if the stressor resolves
66	longstanding affective instability (rapid fluctuation of mood including anger, low mood and euphoria)
67	Stress-related paranoia or dissociations
68	Recurrent suicidal or self-harm behaviour (cutting, overdosing)
69	Unstable relationships or fear of abandonment
70	Pervasive hyperavctivity, impulsivity or inattention started before age 7 years

consisting of a treating psychiatrist from both the psychiatric hospital and outpatient clinics. In order to minimise selection bias, all the patient records that were available in the respective psychiatric wards on the day of data collection were included. Similarly, all new diagnostic assessments of the outpatient clinics for the chosen data collection time periods were included. For each patient clinical features including screening symptoms that were present, were extracted from their records and used as input to the DLD algorithm. The diagnoses produced by the algorithm were compared with the actual diagnoses given by the treating team, and the results at $t_P = t_c = 0.5$ are shown in Fig. 4.

It is important to note that the DLD algorithm did not produce any false negatives, which can have serious implications because false negatives have the effect that an actual diagnosis is missed. The DLD algorithm can, however, produce false positives for diagnoses whose diagnostic criteria is a subset within the diagnostic criteria of actual diagnosis. This is because of limitations of the orthogonal vector projection method in its current form. However, every time such a false positive diagnosis was made, the actual diagnosis was also made and listed with its likelihood value. For example, for every patient with the diagnosis of schizophrenia, the DLD algorithm produces false positive for delusional disorder. Similarly, false positives were produced for the diagnosis of major depressive episode when there were a true diagnoses of Major depressive disorder recurrent and bipolar affective disorder. Since these diagnoses are closely related and with the same group, it is clinically valid to consider them as differential diagnoses, and any harmful consequences of such false positives are minimal since the true diagnosis was simultaneously made.

The results were further analysed by grouping the related diagnoses into main diagnostic categories as in DSM-V, and the sensitivity and specificity, which are standard measures used in clinical medicine to assess accuracy of diagnostic

Table 2 Knowledgebase with row and column numbers corresponding to index of clinical features, including screening symptoms and diagnoses respectively

Index	1	2	3	4	5	6	7	8	9	10	11	12	13	14	15	16	17	18	19	20	21	22	23	24	25	26	27	28	29	30	31	32	33	34	35	36	37	38	39	40	41	42	43	44
1	0,	0,	0,	0,	0,	0,	0,	0,	0,	0,	0,	0,	0,	0,	0,	0,	0,	0,	1,	0,	1,	1,	1,	1,	0,	0,	0,	0,	1,	0,	0,	0,	0,	0,	0,	0,	0,	1,	1,	1,	1,	0,	0,	0,
2	0,	0,	0,	0,	0,	0,	0,	0,	0,	0,	0,	0,	0,	0,	0,	0,	0,	0,	0,	1,	1,	0,	1,	0,	1,	0,	0,	0,	0,	0,	0,	0,	0,	0,	0,	0,	0,	0,	1,	0,	0,	1,	1,	0,
3	0,	0,	0,	0,	0,	0,	0,	0,	0,	0,	0,	0,	0,	0,	0,	0,	0,	0,	1,	0,	1,	1,	0,	0,	0,	0,	0,	0,	0,	0,	0,	0,	0,	0,	0,	0,	0,	1,	1,	0,	0,	0,	0,	0,
4	0,	0,	0,	0,	0,	0,	0,	0,	0,	0,	0,	0,	0,	0,	0,	0,	0,	0,	0,	0,	0,	0,	0,	0,	0,	1,	1,	1,	1,	1,	0,	0,	0,	0,	0,	0,	0,	1,	1,	0,	0,	0,	0,	0,
5	1,	1,	0,	0,	0,	0,	0,	0,	0,	0,	0,	0,	0,	1,	1,	1,	0,	0,	0,	1,	0,	0,	1,	0,	1,	0,	0,	1,	0,	0,	0,	0,	0,	0,	0,	0,	0,	0,	0,	0,	0,	0,	0,	1,
6	1,	0,	0,	0,	0,	0,	0,	0,	0,	0,	0,	0,	0,	0,	0,	0,	0,	0,	0,	0,	0,	0,	0,	0,	1,	0,	0,	0,	0,	0,	0,	0,	0,	0,	0,	0,	0,	0,	1,	0,	0,	0,	0,	0,
7	0,	0,	0,	0,	0,	0,	0,	0,	0,	0,	0,	0,	0,	1,	1,	1,	1,	1,	0,	0,	0,	0,	0,	0,	0,	0,	0,	0,	0,	0,	0,	0,	0,	0,	0,	0,	0,	0,	0,	0,	0,	0,	0,	0,
8	0,	1,	1,	1,	1,	1,	0,	0,	0,	0,	0,	0,	0,	0,	0,	0,	0,	0,	0,	0,	0,	0,	0,	0,	0,	0,	0,	0,	0,	0,	0,	0,	0,	0,	0,	0,	0,	0,	0,	0,	0,	0,	0,	0,
9	0,	0,	0,	0,	0,	0,	0,	0,	0,	0,	0,	0,	0,	0,	0,	0,	0,	0,	0,	0,	0,	0,	0,	0,	0,	0,	0,	0,	0,	0,	1,	1,	1,	1,	1,	0,	0,	0,	0,	0,	0,	0,	0,	0,
10	0,	0,	0,	0,	0,	0,	0,	0,	0,	0,	0,	0,	0,	0,	0,	0,	0,	0,	0,	0,	0,	0,	0,	0,	0,	0,	0,	0,	0,	0,	0,	0,	0,	0,	0,	1,	1,	0,	1,	0,	0,	0,	0,	0,
11	0,	0,	0,	0,	1,	1,	1,	1,	1,	1,	1,	1,	1,	0,	0,	0,	1,	1,	0,	0,	0,	0,	0,	0,	0,	0,	0,	0,	0,	0,	0,	0,	0,	0,	0,	0,	0,	0,	0,	1,	1,	1,	1,	0,
12	1,	1,	0,	0,	0,	0,	0,	0,	0,	0,	0,	0,	0,	0,	0,	0,	0,	0,	0,	0,	0,	0,	0,	0,	0,	0,	0,	0,	0,	0,	0,	0,	0,	0,	0,	0,	0,	0,	0,	0,	0,	0,	0,	0,
13	5,	0,	0,	0,	0,	0,	0,	0,	0,	0,	0,	0,	0,	0,	0,	0,	0,	0,	0,	0,	0,	0,	0,	0,	0,	0,	0,	0,	0,	0,	0,	0,	0,	0,	0,	0,	0,	0,	0,	0,	0,	0,	0,	0,
14	1,	1,	0,	0,	0,	0,	0,	0,	0,	0,	0,	0,	0,	0,	0,	0,	0,	0,	0,	0,	0,	0,	0,	0,	0,	0,	0,	0,	0,	0,	0,	0,	0,	0,	0,	0,	0,	0,	0,	0,	0,	0,	0,	0,
15	0,	0,	1,	1,	1,	0,	0,	0,	0,	0,	0,	0,	0,	0,	0,	0,	0,	0,	0,	0,	0,	0,	0,	0,	0,	0,	0,	0,	0,	0,	0,	0,	0,	0,	0,	0,	0,	0,	0,	0,	0,	0,	0,	0,
16	0,	0,	0,	1,	1,	0,	0,	0,	0,	0,	0,	0,	0,	0,	0,	0,	0,	0,	0,	0,	0,	0,	0,	0,	0,	0,	0,	0,	0,	0,	0,	0,	0,	0,	0,	0,	0,	0,	0,	0,	0,	0,	0,	0,
17	0,	0,	1,	0,	0,	0,	0,	0,	0,	0,	0,	0,	0,	0,	0,	0,	0,	0,	0,	0,	0,	0,	0,	0,	0,	0,	0,	0,	0,	0,	0,	0,	0,	0,	0,	0,	0,	0,	0,	0,	0,	0,	0,	0,
18	0,	0,	0,	0,	3,	0,	0,	0,	0,	0,	0,	0,	0,	0,	0,	0,	0,	0,	0,	0,	0,	0,	0,	0,	0,	0,	0,	0,	0,	0,	0,	0,	0,	0,	0,	0,	0,	0,	0,	0,	0,	0,	0,	0,
19	0,	0,	3,	0,	0,	0,	0,	0,	0,	0,	0,	0,	0,	0,	0,	0,	0,	0,	0,	0,	0,	0,	0,	0,	0,	0,	0,	0,	0,	0,	0,	0,	0,	0,	0,	0,	0,	0,	0,	0,	0,	0,	0,	0,
20	0,	0,	0,	0,	0,	2,	0,	0,	0,	0,	0,	0,	0,	0,	0,	0,	0,	0,	0,	0,	0,	0,	0,	0,	0,	0,	0,	0,	0,	0,	0,	0,	0,	0,	0,	0,	0,	0,	0,	0,	3,	0,	0,	0,
21	0,	0,	0,	0,	0,	1,	0,	1,	0,	0,	0,	0,	0,	0,	0,	0,	0,	0,	0,	0,	0,	0,	0,	0,	0,	0,	0,	0,	0,	0,	0,	0,	0,	0,	0,	0,	0,	0,	0,	0,	3,	0,	0,	0,
22	0,	0,	0,	0,	0,	2,	0,	0,	0,	0,	0,	0,	0,	0,	0,	0,	0,	0,	0,	0,	0,	0,	0,	0,	0,	0,	0,	0,	0,	0,	0,	0,	0,	0,	0,	0,	0,	0,	0,	0,	3,	0,	0,	0,
23	0,	0,	0,	0,	0,	0,	1,	0,	0,	0,	0,	0,	0,	0,	0,	0,	0,	0,	0,	0,	0,	0,	0,	0,	0,	0,	0,	0,	0,	0,	0,	0,	0,	0,	0,	0,	0,	0,	0,	0,	0,	0,	0,	0,
24	0,	0,	0,	0,	0,	0,	0,	0,	3,	0,	0,	0,	0,	0,	0,	0,	0,	0,	0,	0,	0,	0,	0,	0,	0,	0,	0,	0,	0,	0,	0,	0,	0,	0,	0,	0,	0,	0,	0,	0,	0,	0,	0,	0,
25	0,	0,	0,	0,	0,	0,	0,	0,	1,	1,	0,	0,	0,	0,	0,	0,	0,	0,	0,	0,	0,	0,	0,	0,	0,	0,	0,	0,	0,	0,	0,	0,	0,	0,	0,	0,	0,	0,	0,	0,	0,	0,	0,	0,
26	0,	0,	0,	0,	0,	0,	0,	0,	0,	0,	0,	0,	3,	0,	0,	0,	0,	0,	0,	0,	0,	0,	0,	0,	0,	0,	0,	0,	0,	0,	0,	0,	0,	0,	0,	0,	0,	0,	0,	0,	0,	0,	0,	0,
27	0,	0,	0,	0,	0,	0,	0,	0,	0,	0,	0,	0,	1,	1,	0,	0,	0,	0,	0,	0,	0,	0,	0,	0,	0,	0,	0,	0,	0,	0,	0,	0,	0,	0,	0,	0,	0,	0,	0,	0,	0,	0,	0,	0,
28	0,	0,	0,	0,	0,	0,	0,	0,	0,	0,	0,	0,	3,	0,	0,	0,	0,	0,	0,	0,	0,	0,	0,	0,	0,	0,	0,	0,	0,	0,	0,	0,	0,	0,	0,	0,	0,	0,	0,	0,	0,	0,	0,	0,
29	0,	0,	0,	0,	0,	0,	0,	0,	0,	0,	4,	0,	0,	0,	0,	0,	0,	0,	0,	0,	0,	3,	0,	0,	0,	0,	0,	0,	0,	0,	0,	0,	0,	0,	0,	0,	0,	0,	0,	0,	0,	3,	3,	0,
30	0,	0,	0,	0,	0,	0,	0,	0,	0,	0,	1,	1,	0,	0,	0,	0,	0,	0,	0,	0,	0,	0,	0,	0,	0,	0,	0,	0,	0,	0,	0,	0,	0,	0,	0,	0,	0,	0,	0,	0,	0,	3,	3,	0,
31	0,	0,	0,	1,	0,	0,	0,	0,	0,	0,	0,	0,	0,	0,	1,	0,	1,	1,	1,	0,	0,	0,	0,	0,	0,	0,	0,	0,	0,	0,	0,	0,	0,	0,	0,	0,	0,	0,	0,	0,	0,	0,	0,	0,
32	0,	0,	0,	0,	0,	0,	0,	0,	0,	0,	0,	0,	0,	0,	5,	0,	0,	1,	1,	0,	0,	0,	0,	0,	0,	0,	0,	0,	0,	0,	0,	0,	0,	0,	0,	0,	0,	0,	0,	0,	0,	0,	0,	0,
33	0,	0,	0,	0,	0,	0,	0,	0,	0,	0,	0,	0,	0,	0,	1,	0,	0,	0,	0,	0,	0,	0,	0,	0,	0,	0,	0,	0,	0,	0,	0,	0,	0,	0,	0,	0,	0,	0,	0,	0,	0,	0,	0,	0,
34	0,	0,	0,	0,	0,	0,	0,	0,	0,	0,	0,	0,	0,	0,	2,	0,	0,	0,	0,	0,	0,	0,	0,	0,	0,	0,	0,	0,	0,	0,	0,	0,	0,	0,	0,	0,	0,	0,	0,	0,	0,	0,	0,	0,
35	0,	0,	0,	0,	0,	0,	0,	0,	0,	0,	0,	0,	0,	0,	0,	1,	0,	0,	0,	0,	0,	0,	0,	0,	0,	0,	0,	0,	0,	0,	0,	0,	0,	0,	0,	0,	0,	0,	0,	0,	0,	0,	0,	0,
36	0,	0,	0,	0,	0,	0,	0,	0,	0,	0,	0,	0,	0,	0,	0,	0,	0,	5,	0,	0,	0,	0,	0,	0,	0,	0,	0,	0,	0,	0,	0,	0,	0,	0,	0,	0,	0,	0,	0,	0,	0,	0,	0,	0,
37	0,	0,	0,	0,	0,	0,	0,	0,	0,	0,	0,	0,	0,	0,	0,	0,	0,	0,	5,	0,	0,	0,	0,	0,	0,	0,	0,	0,	0,	0,	0,	0,	0,	0,	0,	0,	0,	0,	0,	0,	0,	0,	0,	0,
38	0,	0,	0,	0,	0,	0,	0,	0,	0,	0,	0,	0,	0,	0,	0,	0,	0,	0,	0,	3,	0,	3,	0,	1,	0,	0,	0,	0,	0,	0,	0,	0,	0,	0,	0,	0,	0,	0,	0,	0,	1,	1,	0,	0,
39	0,	0,	0,	0,	0,	0,	0,	0,	0,	0,	0,	0,	0,	0,	0,	0,	0,	0,	0,	0,	0,	0,	3,	0,	0,	0,	0,	0,	0,	0,	0,	0,	0,	0,	0,	0,	0,	0,	0,	0,	0,	0,	0,	0,
40	0,	0,	0,	0,	0,	0,	0,	0,	0,	0,	0,	0,	0,	0,	0,	0,	0,	0,	0,	1,	0,	1,	1,	1,	0,	0,	0,	0,	0,	0,	0,	0,	0,	0,	0,	0,	0,	0,	0,	0,	1,	1,	0,	0,
41	0,	0,	0,	0,	0,	0,	0,	0,	0,	0,	0,	0,	0,	0,	0,	0,	0,	0,	0,	1,	0,	1,	1,	1,	0,	0,	0,	0,	0,	0,	0,	0,	0,	0,	0,	0,	0,	0,	0,	0,	1,	1,	0,	0,
42	0,	0,	0,	0,	0,	0,	0,	0,	0,	0,	0,	0,	0,	0,	0,	0,	0,	0,	0,	1,	0,	1,	0,	1,	0,	0,	0,	0,	0,	0,	0,	0,	0,	0,	0,	0,	0,	0,	0,	0,	0,	0,	0,	0,
43	0,	0,	0,	0,	0,	0,	0,	0,	0,	0,	0,	0,	0,	0,	0,	0,	0,	0,	0,	0,	0,	5,	0,	1,	1,	0,	0,	0,	0,	0,	0,	0,	0,	0,	0,	0,	0,	0,	0,	0,	0,	0,	0,	0,
44	0,	0,	0,	0,	0,	0,	0,	0,	0,	0,	0,	0,	0,	0,	0,	0,	0,	0,	0,	0,	1,	0,	0,	0,	1,	0,	0,	0,	0,	0,	0,	0,	0,	0,	0,	0,	0,	0,	0,	0,	0,	0,	1,	0,
45	0,	0,	0,	0,	0,	0,	0,	0,	0,	0,	0,	0,	0,	0,	0,	0,	0,	0,	0,	0,	1,	0,	0,	0,	1,	0,	0,	0,	0,	0,	0,	0,	0,	0,	0,	0,	0,	0,	0,	0,	0,	0,	1,	0,
46	0,	0,	0,	0,	0,	0,	0,	0,	0,	0,	0,	0,	0,	0,	0,	0,	0,	0,	0,	0,	1,	0,	0,	0,	1,	0,	0,	0,	0,	0,	0,	0,	0,	0,	0,	0,	0,	0,	0,	0,	0,	0,	1,	0,
47	0,	0,	0,	0,	0,	0,	0,	0,	0,	0,	0,	0,	0,	0,	0,	0,	0,	0,	0,	0,	1,	0,	0,	0,	1,	0,	0,	0,	0,	0,	0,	0,	0,	0,	0,	0,	0,	0,	0,	0,	0,	0,	1,	0,
48	0,	0,	0,	0,	0,	0,	0,	0,	0,	0,	0,	0,	0,	0,	0,	0,	0,	0,	0,	0,	0,	0,	0,	4,	4,	0,	0,	0,	0,	0,	0,	0,	0,	0,	0,	0,	0,	0,	0,	0,	0,	0,	0,	0,
49	0,	0,	0,	0,	0,	0,	0,	0,	0,	0,	0,	0,	0,	0,	0,	0,	0,	0,	0,	0,	0,	0,	0,	0,	0,	0,	1,	0,	0,	0,	0,	0,	0,	0,	0,	0,	0,	0,	0,	0,	0,	0,	0,	0,
50	0,	0,	0,	0,	0,	0,	0,	0,	0,	0,	0,	0,	0,	0,	0,	0,	0,	0,	0,	0,	0,	0,	0,	0,	0,	1,	0,	0,	0,	0,	0,	0,	0,	0,	0,	0,	0,	0,	0,	0,	0,	0,	0,	0,
51	0,	0,	0,	0,	0,	0,	0,	0,	0,	0,	0,	0,	0,	0,	0,	0,	0,	0,	0,	0,	0,	0,	0,	0,	0,	0,	0,	1,	0,	0,	0,	0,	0,	0,	0,	0,	0,	0,	0,	0,	0,	0,	0,	0,
52	0,	0,	0,	0,	0,	0,	0,	0,	0,	0,	0,	0,	0,	0,	0,	0,	0,	0,	0,	0,	0,	0,	0,	0,	0,	0,	0,	0,	1,	0,	0,	0,	0,	0,	0,	0,	0,	0,	0,	0,	0,	0,	0,	0,
53	0,	0,	0,	0,	0,	0,	0,	0,	0,	0,	0,	0,	0,	0,	0,	0,	0,	0,	0,	0,	0,	0,	0,	0,	0,	0,	0,	0,	0,	1,	0,	0,	0,	0,	0,	0,	0,	0,	0,	0,	0,	0,	0,	0,
54	0,	0,	0,	0,	0,	0,	0,	0,	0,	0,	0,	0,	0,	0,	0,	0,	0,	0,	0,	0,	0,	0,	0,	0,	0,	0,	0,	0,	0,	1,	0,	0,	0,	0,	0,	0,	0,	0,	0,	0,	0,	0,	0,	0,
55	0,	0,	0,	0,	0,	0,	0,	0,	0,	0,	0,	0,	0,	0,	0,	0,	0,	0,	0,	0,	0,	0,	0,	0,	0,	0,	0,	0,	0,	0,	1,	0,	0,	0,	0,	0,	0,	0,	0,	0,	0,	0,	0,	0,
56	0,	0,	0,	0,	0,	0,	0,	0,	0,	0,	0,	0,	0,	0,	0,	0,	0,	0,	0,	0,	0,	0,	0,	0,	0,	0,	0,	0,	0,	0,	1,	0,	0,	0,	0,	0,	0,	0,	0,	0,	0,	0,	0,	0,
57	0,	0,	0,	0,	0,	0,	0,	0,	0,	0,	0,	0,	0,	0,	0,	0,	0,	0,	0,	0,	0,	0,	0,	0,	0,	0,	0,	0,	0,	0,	0,	1,	0,	0,	0,	0,	0,	0,	0,	0,	0,	0,	0,	0,
58	0,	0,	0,	0,	0,	0,	0,	0,	0,	0,	0,	0,	0,	0,	0,	0,	0,	0,	0,	0,	0,	0,	0,	0,	0,	0,	0,	0,	0,	0,	0,	0,	1,	0,	0,	0,	0,	0,	0,	0,	0,	0,	0,	0,
59	0,	0,	0,	0,	0,	0,	0,	0,	0,	0,	0,	0,	0,	0,	0,	0,	0,	0,	0,	0,	0,	0,	0,	0,	0,	0,	0,	0,	0,	0,	0,	0,	0,	1,	0,	0,	0,	0,	0,	0,	0,	0,	0,	0,
60	0,	0,	0,	0,	0,	0,	0,	0,	0,	0,	0,	0,	0,	0,	0,	0,	0,	0,	0,	0,	0,	0,	0,	0,	0,	0,	0,	0,	0,	0,	0,	0,	0,	0,	1,	0,	0,	0,	0,	0,	0,	0,	0,	0,
61	0,	0,	0,	0,	0,	0,	0,	0,	0,	0,	0,	0,	0,	0,	0,	0,	0,	0,	0,	0,	0,	0,	0,	0,	0,	0,	0,	0,	0,	0,	0,	0,	0,	0,	0,	1,	0,	0,	0,	0,	0,	0,	0,	0,
62	0,	0,	0,	0,	0,	0,	0,	0,	0,	0,	0,	0,	0,	0,	0,	0,	0,	0,	0,	0,	0,	0,	0,	0,	0,	0,	0,	0,	0,	0,	0,	0,	0,	0,	0,	0,	1,	0,	0,	0,	0,	0,	0,	0,
63	0,	0,	0,	0,	0,	0,	0,	0,	0,	0,	0,	0,	0,	0,	0,	0,	0,	0,	0,	0,	0,	0,	0,	0,	0,	0,	0,	0,	0,	0,	0,	0,	0,	0,	0,	0,	1,	0,	0,	0,	0,	0,	0,	0,
64	0,	0,	0,	0,	0,	0,	0,	0,	0,	0,	0,	0,	0,	0,	0,	0,	0,	0,	0,	0,	0,	0,	0,	0,	0,	0,	0,	0,	0,	0,	0,	0,	0,	0,	0,	0,	0,	1,	0,	0,	0,	0,	0,	0,
65	0,	0,	0,	0,	0,	0,	0,	0,	0,	0,	0,	0,	0,	0,	0,	0,	0,	0,	0,	0,	0,	0,	0,	0,	0,	0,	0,	0,	0,	0,	0,	0,	0,	0,	0,	0,	0,	0,	1,	0,	0,	0,	0,	0,
66	0,	0,	0,	0,	0,	0,	0,	0,	0,	0,	0,	0,	0,	0,	0,	0,	0,	0,	0,	0,	0,	0,	0,	0,	0,	0,	0,	0,	0,	0,	0,	0,	0,	0,	0,	0,	0,	0,	0,	1,	0,	0,	0,	0,
67	0,	0,	0,	0,	0,	0,	0,	0,	0,	0,	0,	0,	0,	0,	0,	0,	0,	0,	0,	0,	0,	0,	0,	0,	0,	0,	0,	0,	0,	0,	0,	0,	0,	0,	0,	0,	0,	0,	0,	1,	0,	0,	0,	0,
68	0,	0,	0,	0,	0,	0,	0,	0,	0,	0,	0,	0,	0,	0,	0,	0,	0,	0,	0,	0,	0,	0,	0,	0,	0,	0,	0,	0,	0,	0,	0,	0,	0,	0,	0,	0,	0,	0,	0,	1,	0,	0,	0,	0,
69	0,	0,	0,	0,	0,	0,	0,	0,	0,	0,	0,	0,	0,	0,	0,	0,	0,	0,	0,	0,	0,	0,	0,	0,	0,	0,	0,	0,	0,	0,	0,	0,	0,	0,	0,	0,	0,	0,	0,	1,	0,	0,	0,	0,
70	0,	0,	0,	0,	0,	0,	0,	0,	0,	0,	0,	0,	0,	0,	0,	0,	0,	0,	0,	0,	0,	0,	0,	0,	0,	0,	0,	0,	0,	0,	0,	0,	0,	0,	0,	0,	0,	0,	0,	0,	0,	0,	0,	4,

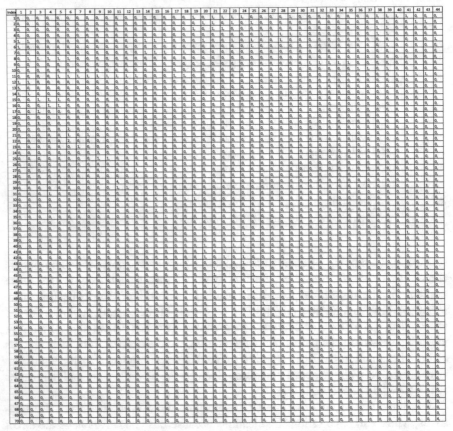

instruments [27], for each diagnostic category are shown in the table in Fig. 3. When the related disorders were considered as a group, false positives were reduced to only 2 for psychotic disorders and 3 for mood disorders. Anxiety disorders as a group did not have any false positives.

This paper has presented an improved version of the ST algorithm, which can potentially be used to develop large scale medical expert systems that cover a broader clinical knowledge across various medical specialities. However, achieving such a large scale medical expert system is a challenging task, which requires large-scale resources and manpower. Also, it requires integration of various technologies such as image processing, speech recognition, and biomedical signal processing depending on how sophisticatedly the abstraction step of the algorithm is implemented. Unless, it can be proven that the system is achievable, such a large-scale effort would a waste. The existence of ST algorithm provides the

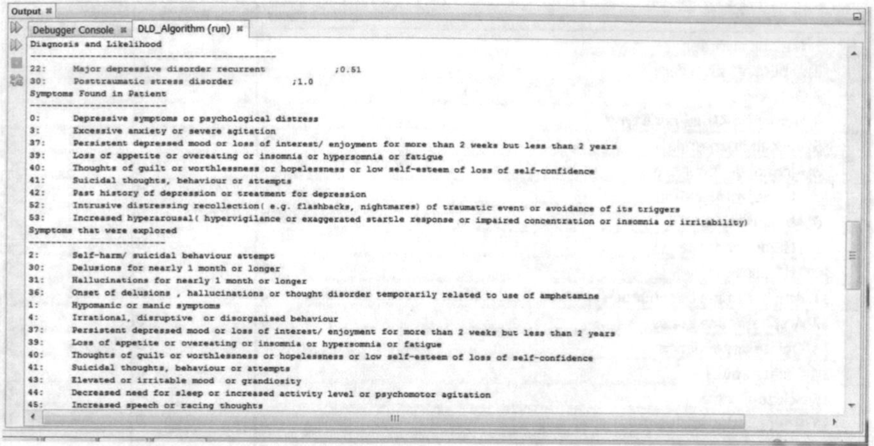

Fig. 2 Netbean implementation of the algorithm using java

Disorder group	Sensitivity(95% CI)	Specificity(95% CI)
Psychotic disorders (18-22)	100(84.56 to 100)%	98.48(94.63 to 99.82)%
Mood disorders(20-25,39, 40-43)	100(94.4% to 100)%	96.74(90.77% to 99.32)%
Anxiety disorders(26-31)	100(92.13% to 100)%	100(96.67% to 100)%

Fig. 3 Sensitivity and specificity analysis for major diagnostic categories. Conclusion

required theoretical proof that such systems are achievable. Whilst a simpler design of knowledgebase with two layers (i.e. clinical features and diagnoses) is presented in this paper, it is possible to enhance it by adding more layers(e.g. a layer of attributes that are associated with each clinical feature) as required.

5 Conclusion

This paper described the design, implementation, and evaluation of the DLD algorithm as applied to clinical psychiatry. Even though the DLD algorithm is not able to perform an exhaustive search compared to its rival ST algorithm, the results using actual patient data indicated a comparable accuracy, which is significant given its relative efficiency ($O\left(n^2\right)$ compare to $O\left(n^3\right)$ of ST algorithm). Even though the risk of making false negative diagnoses can be minimised by carefully choosing screening symptoms in such a way that they cover broader diagnostic categories, the risk of false negatives cannot be eliminated entirely. However, because of its relative efficiency, the DLD algorithm can ideally be implemented as a triaging tool to make a preliminary diagnosis, whilst the ST algorithm can be used for excluding important diagnoses via its exhaustive search. The use of both

Index	Diagnosis	Number of cases	False Positives	False Negatives
1	Autistic disorder	0	0	0
2	Asperger's disorder	1	0	0
3	Delirium	1	0	0
4	Dementia Alzheimer's type	2	2	0
5	Vascular dementia	1	1	0
6	Alcohol dependence	4	0	0
7	Alcohol intoxication	0	0	0
8	Alcohol abuse	2	3	0
9	THC dependence	5	0	0
10	THC abuse	1	4	0
11	Amphetamine dependence	2	0	0
12	Amphetamine abuse	0	0	0
13	Opiate dependence	0	0	0
14	Opiate abuse	0	0	0
15	Schizophrenia	17	4	0
16	Brief psychotic disorder	2	0	0
17	Delusional disorder	3	16	0
18	THC induced psychotic disorder	0	1	0
19	Amphetamine induced psychotic disorder	0	0	0
20	Major depressive episode	15	25	0
21	Manic episode	2	5	0
22	Major depressive disorder recurrent	23	5	0
23	Dysthymic disorder	1	4	0
24	Bipolar disorder currently depressed	3	0	0
25	Bipolar disorder currently manic	6	0	0
26	Agoraphobia	5	0	0
27	Panic disorder	8	0	0
28	Social anxiety disorder	11	0	0
29	Obsessive compulsive disorder	5	0	0
30	Posttraumatic stress disorder	7	1	0
31	Generalised anxiety disorder	9	0	0
32	Somatization disorder	0	0	0
33	Conversion disorder	0	0	0
34	Hypochondriasis	0	0	0
35	Body dysmorphic disorder	0	0	0
36	Factitious disorder	0	0	0
37	Anorexia nervosa	1	0	0
38	Bulimia nervosa	0	0	0
39	Adjustment disorder	8	0	0
40	Borderline personality disorder	4	2	0
41	Alcohol induced depressive disorder	0	3	0
42	Amphetamine induced depressive disorder	1	0	0
43	Amphetamine induced hypomanic/manic symptoms	0	0	0
44	Attention deficit hyperactivity disorder	1	0	0
45	No diagnosis	3	0	0

Fig. 4 Results of the DLD algorithm

algorithms, along with the orthogonal vector projection method, provides a reasonable theoretical foundation for automating medical diagnostic reasoning. It should be noted that, while both algorithms in their current form have successfully simplified the abstraction of clinical features displayed by patients, they still very much rely on involvement of a clinician. It is the purpose of future research to aim at achieving complete automation of abstraction in order to develop fully automated medical expert systems. Also, the orthogonal vector projection method needs to be improved in order to minimise the occurrence of false positives that result in listing of potential diagnoses whose diagnostic criteria form a subset within the diagnostic criteria of one or more of the actual diagnoses.

Appendix

See Tables 1 and 2.

References

1. Shortliffe, E.H., Buchanan, B.G.: A model of inexact reasoning in medicine. Math. Biosci. **23**, 351–379, 4 (1975)
2. Reggia, J.A., Peng, Y.: Modeling diagnostic reasoning: a summary of parsimonious covering theory. Comput. Methods Programs Biomed. **25**, 125–134 (1987)
3. Stausberg, J.R., Person, M.: A process model of diagnostic reasoning in medicine. Int. J. Med. Inform. **54**, 9–23 (1999)
4. Wortman, P.M.: Medical diagnosis: an information-processing approach. Comput. Biomed. Res. **5**, 315–328 (1972)
5. Mandin, H., Jones, A., Woloschuk, W., Harasym, P.: Helping students learn to think like experts when solving clinical problems. Acad. Med. **72**, 173–179 (1997)
6. Hunt, E.: Cognitive science: definition, status, and questions. Annu. Rev. Psychol. **40**, 603–629 (1989)
7. Norman, G.R., Coblentz, C.L., Brooks, L.R., Babcook, C.J.: Expertise in visual diagnosis—a review of the literature. Acad. Med. **66**(suppl), s78–s83 (1992)
8. Elstein, A.S., Shulman, L.S., Sprafka, S.A.: Medical Problem-Solving: an Analysis of Clinical Reasoning: Cambridge. Harvard University Press, Cambridge, MA (1978)
9. Andreassen, S., Jensen, F.V., Olesen, K.G.: Medical expert systems based on causal probabilistic networks. Int. J. Bio-Med. Comput. **28**, 1–30, 5 (1991)
10. Chard, T., Rubenstein, E.M.: A model-based system to determine the relative value of different variables in a diagnostic system using Bayes theorem. Int. J. Bio-Med. Comput. **24**, 133–142, 7 (1989)
11. Todd, B.S., Stamper, R., Macpherson, P.: A probabilistic rule-based expert system. Int. J. Bio-Med. Comput. **33**, 129–148, 9 (1993)
12. Boegl, K., Adlassnig, K.-P., Hayashi, Y., Rothenfluh, T.E., Leitich, H.: Knowledge acquisition in the fuzzy knowledge representation framework of a medical consultation system. Artif. Intell. Med. **30**, 1–26, 1 (2004)
13. Godo, L.S., de Mántaras, R.L., Puyol-Gruart, J., Sierra, C.: Renoir, Pneumon-IA and Terap-IA: three medical applications based on fuzzy logic. Artif. Intell. Med. **21**, 153–162, 1 (2001)
14. Vetterlein, T., Ciabattoni, A.: On the (fuzzy) logical content of CADIAG-2. Fuzzy Sets Syst. **161**, 1941–1958. Accessed 16 July 2010
15. Wolfram, D.A.: An appraisal of INTERNIST-I. Artif. Intell. Med. **7**, 93–116 (1995)
16. Wears, R.L., Berg, M.: Computer technology and clinical work still waiting for Godot. JAMA **293**, 1261–1263 (2005)

17. Das, A.K.: Computers in psychiatry: a review of past programs and an analysis of historical trends. Psychiatr. Q. **73**, 351–365 (2002)
18. Ramoni, M., Stefanelli, M., Magnani, L., Barosi, G.: An epistemological framework for medical knowledge-based systems. IEEE Trans. Syst. Man Cybern. **22**, 1361–1375 (1992)
19. Peirce, C.S.: Illustrations of the logic of science, sixth paper-deduction, induction, hypothesis. Popular Sci. Mon. **1**, 470–482 (1878)
20. Fernando, I., Henskens, F.: A modified case-based reasoning approach for triaging psychiatric patients using a similarity measure derived from orthogonal vector projection. In: Chalup, S., Blair, A., Randall, M. (eds.) Artificial Life and Computational Intelligence, vol. 8955, pp. 360–372. Springer International Publishing (2015)
21. Fernando, I., Henskens, F.: Select and Test (ST) algorithm for medical diagnostic reasoning. In: SNPD 2016. Shanghai, China (2016)
22. Fernando, I., Henskens, F.: Select and test algorithm for inference in medical diag-nostic reasoning: implementation and evaluation in clinical psychiatry. In: ICIS 2016. Okayama, Japan (2016) (submitted)
23. American Psychiatric Association, Diagnostic and Statistical Manual of Mental Disorders: Dsm-5: Amer Psychiatric Pub Incorporated (2013)
24. W. H. Organization: The ICD-10 Classification of Mental and Behavioural Disorders: Clinical Descriptions and Diagnostic Guidelines: Royal College of Psychiatrists (1992)
25. Fernando, I., Cohen, M., Henskens, F.: A systematic approach to clinical reasoning in psychiatry. Australas. Psychiatry **21**, 224–230 (2013)
26. Fernando, I., Henskens, F.: Drill-locate-drill algorithm for diagnostic reasoning in psychiatry. Int. J. Mach. Learn. Comput. **3**, 449–452 (2013)
27. Altman, D.G., Bland, J.M.: Diagnostic tests. 1: sensitivity and specificity. BMJ Br. Med. J. **308,** 1552 (1994)

Implementation of Artificial Neural Network and Multilevel of Discrete Wavelet Transform for Voice Recognition

Bandhit Suksiri and Masahiro Fukumoto

Abstract This paper presents an implementation of simple Artificial Neural Network model and multilevel of Discrete Wavelet Transform as feature extractions, which is achieved to increase the high recognition rates up to 95 % instead of Short-time Fourier Transform in the conversation background noises at noises up to 65 dB. The performance evaluation has been demonstrated in terms of correct recognition rate, maximum noise power of interfering sounds, hit rates, false alarm rates and miss rates. The proposed method offers a potential alternative to intelligence voice recognition system in speech analysis-synthesis and recognition applications.

Keywords Discrete wavelet transform • Voice recognition • Artificial neural network • Feature extractions

1 Introduction

During the past 65 years, voice recognition is being extensively implemented for the classification of sound types. The variety of voice recognition techniques have been developed to increase the efficiency of recognitive accuracy, statistical pattern recognition, signal processing and recognition rates as shown in [1].

According to a lot of research, a number of algorithms have been proposed and suggested as potential solutions to recognize human's speech, i.e., the simply probability distribution fitting methods such as, Structural Maximum A Posteriori,

B. Suksiri (✉)
Graduate School of Engineering, Kochi University of Technology (KUT),
Kami City, Kochi 782-8502, Japan
e-mail: 187001v@gs.kochi-tech.ac.jp

M. Fukumoto (✉)
School of Information, Kochi University of Technology (KUT),
Kami City, Kochi 782-8502, Japan
e-mail: fukumoto.masahiro@kochi-tech.ac.jp

R. Lee (ed.), *Computer and Information Science*,
Studies in Computational Intelligence 656, DOI 10.1007/978-3-319-40171-3_2

Parallel Model Composition and Maximum Likelihood Linear Regression. However, the issue of sequential voice input had been being still unsolved.

Ferguson et al. has proposed Hidden Markov Model (HMM) in order to solve an issue of sequential voice input. HMM was employed double stochastic process using an embedded stochastic function in order to determine the value of the hidden states as shown in [1]. High recognition rates design was essentially required state of the art of architecture in HMM using Gaussian Mixture Model (GMM) as shown in [2, 3]. GMM has been traditionally utilized voice models for voice recognition using two feature extractions, a power logarithm of FFT spectrum in order to create Log-power spectrum feature vectors and Mel-Scale Filter Bank Inverse FFT Dimension Reduction in order to created Mel Frequency Cepstral Coefficient feature vectors. GMM offered high voice recognition rate from 60 to 95 % in a static environment by comparison with other machine learning model such as Support Vector Machine and Dual Penalized Logistic Regression Machine as shown in [2]. Nonetheless, large amounts of computational resource are required in GMM.

Pitch-Cluster-Maps (PCMs) model was proposed by Yoko et al. [4] in order to replace the complex training sets with Binarized Frequency Spectrum resulted from simple codebook sets using Short-time Fourier Transform [5–7]. Vector Quantization Approach method was employed lead suitable Real-Time computation than GMM. Nonetheless, PCMs offered voice recognition rate up to 60 % for 6 sound sources environment under low frequency resolution.

This paper aim to propose, the alternative voice recognition utilize Artificial Neural Network and Multilevel of Discrete Wavelet Transform with 3 main advantages. First, Discrete Wavelet Transform has resolved the low frequency prediction issue in order to increases low frequency prediction. Second, the normal conversation background noise issue resolves in the proposed voice recognition. Last, the proposed voice recognition has been improved recognition rates up to 95 % by comparison with other model.

2 Proposed Voice Recognition

The overview of proposed voice recognitions consisted of feature extraction, feature normalization, machine learning as ANN and decision model which summarized in Fig. 1.

2.1 Feature Extraction

The proposed voice recognition utilized the feature extraction as the pre-processing methods in order to transform the voices or signals to the time-frequency represented data. Three pre-processing methods were implemented for voices feature extraction consisted of Short-time Fourier Transform (STFT) and Discrete Wavelet

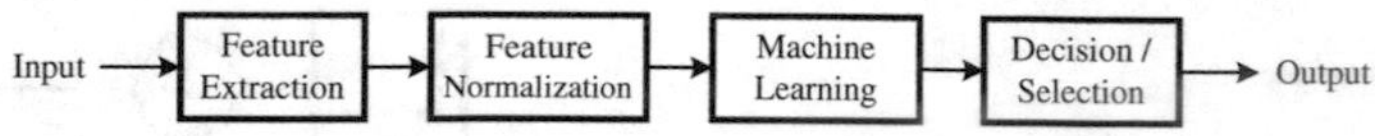

Fig. 1 The proposed voice recognition overview

Transform (DWT) [10]. In general case, Continuous Wavelet Transform (CWT) can be expressed as

$$X_{\Psi}(a,b) = \frac{1}{\sqrt{|a|}} \int_{-\infty}^{\infty} x(t) \cdot \overline{\Psi}\left(\frac{t-b}{a}\right) dt \tag{1}$$

where $\overline{\Psi}(t)$ is the conjugate of Wavelet function, a is input scales which represented as frequency variable, b is input time variable, $x(t)$ is the continuous signal to be transformed and $X_{\Psi}(a,b)$ is the CWT of a complex function represented the magnitude of the continuous signal over time and frequency based on specified Wavelet function.

In particular, DWT transformation decomposes the signal into mutually orthogonal set of wavelets, which is the main difference from the CWT [10] or its implementation for the discrete time series. DWT provides sufficient information in both time and frequency with a significant reduction in the computation time than CWT [10]. DWT can be constructed from convolution of the signal with the impulse response of the filter expressed as

$$\phi[n] = \sum_{i=-\infty}^{\infty} a_i \cdot \phi[Sn - i] \tag{2}$$

where $\phi[n]$ is the dilation reference equation as discrete signals from input to output states, S is a scaling factor to be assign value to 2, n is time index and a_i consists of two scaling functions obtained from each Wavelet function know as Quadrature Mirror Filter.

DWT equation can be represented as a binary hierarchical tree of LPF and HPF, in other words, it can be defined as Filter Banks as shown in Fig. 2. In Filter Banks analysis, lengths of discrete signals are reduced by halved per level. The effect of shifting and scaling process from (1) to (2) produces a time-scale representation as shown in Fig. 3. The graphs show the signal amplitudes in both of time and frequency domain using STFT for left-hand graph and CWT for right-hand graph. The vertical axis is represented frequency band and horizontal axis is represented time domain. It can be seen from a comparison with STFT and CWT, Wavelet Transform offers a superior temporal resolution of time resolution at high frequency components and scale resolution at low frequency components [10], which usually give a voice signal and its main characteristics or identity.

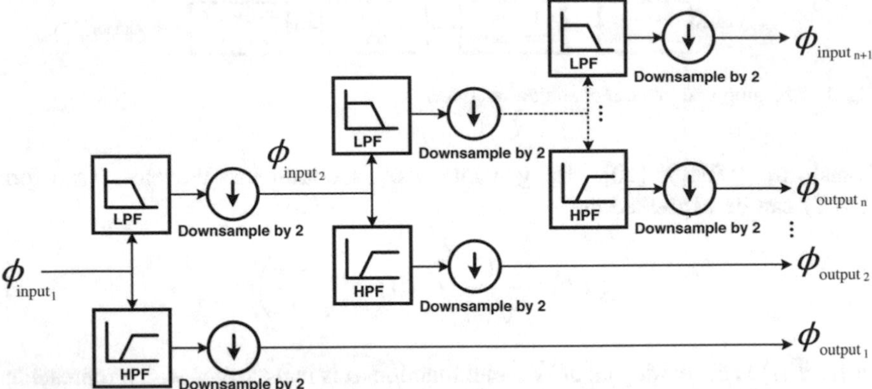

Fig. 2 DWT Filter Bank representation

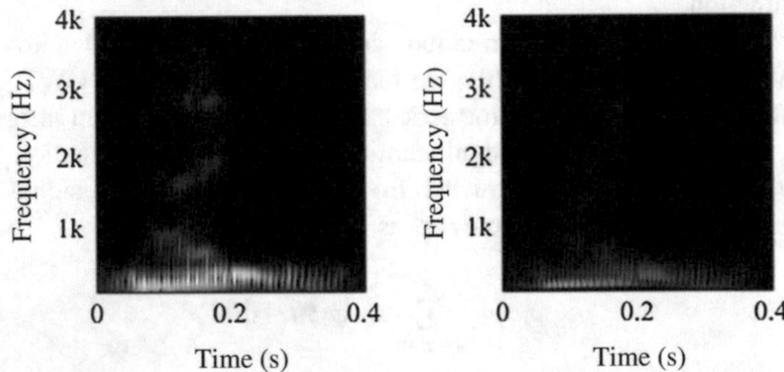

Fig. 3 The comparison of STFT (*left*) and CWT (*right*)

2.2 Feature Normalization

In order to increases the speed convergences of the machine learning algorithm, Feature Normalization method is utilized with simplest form is given as follow

$$\tilde{x} = \frac{x - \min(x)}{\max(x) - \min(x)} - \bar{x} \tag{3}$$

where $\tilde{x}$ is the normalized vector and x is original vector determine from feature extraction and $\bar{x}$ is its offset of signals from zero. Feature Normalization offers the range of original vector to scale the range between 0 and 1.

2.3 Artificial Neural Network Model

Artificial Neuron Network (ANN) [8] is an adaptive system that changes structure based on external and internal information that flows through the network. ANN is considered nonlinear statistical data modeling tools where the complex relationships between inputs and outputs are modeled or patterns are found. Therefore, the proposed voice recognition utilized ANN in order to recognize a characteristics or identity of human speech.

The novel network topology name the nth-order All-features-connecting topology is represented by H_n as illustrated in Fig. 4 where x_f is an input vector in each frequency band which is calculated from feature extraction model and y is a class probability vector which is calculated by ANN. H_n model utilizes the network of A, B and C-class in order to construct simple network topology with those network were shown in Table 1. The four main conditions of the novel network topology are defined. First, numbers of layers are defined from order of H_n where $n > 0$. Second, numbers of input networks are related to number of input time index, i.e., size of

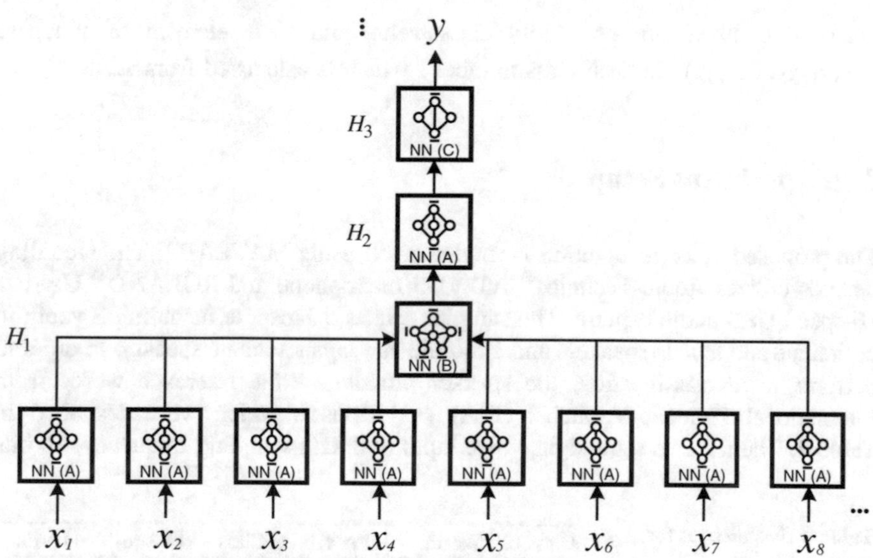

Fig. 4 All-features-connecting topology

Table 1 The Proposed neuron network architectures

Class name	Control input	Control output	Transfer function
A-class	Single	Single	Log-sigmoid
B-class	Multiple	Single	Log-sigmoid
C-class	Single	Single	Softmax

scales vector for CWT and number of levels for DWT. Third, input networks are required the connection of all single outputs to the first block of network series. The input, middle and output networks are A, B and C-class network, respectively, which give as a last condition.

In order to train specified ANN, Scaled Conjugate Gradient Backpropagation [9] supervised learning rule is employed. Additionally, specified ANN utilizes pre-learning rule using Autoassociator [11, 12] to initiate weights approximation of final solution lead to accelerate the convergence of the error Backpropagation learning algorithm and reduce dimension from wavelet packet series.

2.4 Decision Model

The output of ANN is represented as vector of class possibility value base on feature set. The decision model is expressed as maximum of class possibility value

$$c = \underset{i \in \aleph}{\operatorname{argmax}}(y_i) \tag{4}$$

where c is maximum possibility class value and y_i is element of y where $y = (y_1, y_2, \ldots, y_n)^T$ in each class number i which is calculated from ANN.

3 Experiment Setup

The proposed voice recognition is implemented using MATLAB®. The recording devices utilizes Audio-Technica® AT-VD3 microphone and ROLAND® UA-101 Hi-speed USB audio capture. The samples selects 5 Japanese including 2 youth in both male and female speakers and 1 middle-age Japanese male speaker. In order to perform word classification, the speaker pronounces the reference words from International Phonetic Alphabet (IPA) [13] datasets which were described in Table 2. The features set assigns voice input to 8 kHz sampling frequency, 16 bits

Table 2 Features set for experimentation

Class	Words	IPA's	Class	Words	IPA's
1	パン	páɴ	10	雑用	zátˢuzi
2	番	báɴ	11	山	jamá
3	先ず	mázu	12	脈	mjakú
4	太陽	táijo	13	風	kaze
5	段々	daɴdaɴ	14	外套	gaito:
6	通知	tˢu: tˢi	15	医学	ígaku
7	何	náni	16	善意	zéɴi
8	蘭	ɽáɴ	17	鼻	hana
9	数字	su:si	18	わ	wa

data resolution and 8000 sample points. The numbers of features set are 450 elements obtains from reference words in dataset with 5 times repeated. In order to perform the performance evaluation, the experiments are selected 20 % for tests set and 80 % for features set.

4 Experimental Results

The performance evaluation was established in term of correct recognition rate which calculated from the summation of true positive and true negative rates in each class. Moreover, maximum noise power of interfering sounds with nonlinear logarithmic scale defines as follows

$$P_{\text{noise, dB}} = 10\log_{10}\left(\frac{P_{\text{noise}}}{P_{\text{ref}}}\right) \quad (5)$$

where $P_{\text{noise,dB}}$ is noise power level in decibel (dB), P_{noise} is noise power level in watt and P_{ref} is reference power level in watt (W). The experimentation assign P_{ref} is 10^{-12} W as a reference for ambient noise level in order to map voice signal conditions over a spatial regime.

Three experiments were conducted with subject to word classification in order to examine appropriate values of Wavelet and ANN parameters. The first experiment proposed an examination of Wavelet function category and its order using set of static parameters shown in Table 3. The experimental results consisted of three Wavelet functions included Daubechies, Symlet and Coiflet Wavelet function with each order from 1 to 16. It is definitely seen from Table 4 that several Wavelet function achieved word classification with correct recognition rates greater than 80 % and noise power of interfering sounds greater than 50 dB. The Wavelet function was selected by two satisfied conditions, maximum values of noise power of interfering sound and correct recognition rate. Therefore, Daubechies 15 Wavelet function revealed the satisfied maximum values of noise power of interfering sound 65.5 dB and correct recognition rate 96.22 %.

Table 3 The first experimental configuration

Parameter name	Value
Subject	Word classification
Feature extraction method	Discrete Wavelet Transform (DWT)
Wavelet level	6
Wavelet function	*Variable parameter*
Network topology	3rd-order All-features-connecting topology (H_3)
Node size in each layer	{1000, 4000, 1000, 18}

Table 4 The first experimental results

Wavelet function						
Order	Daubechies (db)		Symlet (sym)		Coiflet (coif)	
	$P_{noise,}$ dB	Recognition Rate (%)	$P_{noise,}$ dB	Recognition rate (%)	$P_{noise,}$ dB	Recognition rate (%)
1	24.50	90.44	*None*		64.50	94.89
2	60.38	93.11	0.00	88.44	62.50	93.33
3	63.63	94.00	63.75	94.22	63.50	96.00
4	64.25	92.44	55.50	94.00	42.50	92.00
5	55.13	94.67	63.25	94.67	*None*	
6	67.75	94.89	65.25	96.00	*None*	
7	56.00	93.78	61.00	94.67	*None*	
8	57.50	95.56	35.50	91.11	*None*	
9	34.50	92.44	67.50	94.22	*None*	
10	36.25	93.78	64.25	94.44	*None*	
11	59.00	94.00	0.00	84.00	*None*	
12	61.50	95.56	61.25	95.11	*None*	
13	67.50	95.11	67.75	95.78	*None*	
14	58.00	95.33	65.25	94.67	*None*	
15	**65.50**	**96.22**	27.75	91.33	*None*	
16	61.50	96.88	65.75	94.44	*None*	

However, cost functions of proposed voice recognition were obviously influenced by the effect of Wavelet function, Wavelet level and ANN network topology. Hence, the second experiment was designed to optimize Wavelet level and ANN network topology using set of static parameters as shown in Table 5. It is obviously seen from Table 6 that H_3 model with Wavelet level 4 to 8 achieved word classification with noise power of interfering sounds greater than 60 dB and correct recognition rates greater than 94 %. Hence, H_3 model with Wavelet level 6 was selected with two satisfied conditions criteria, minimizing computation and verify the validity inside the ROI in human speech frequency form 130 to 4 kHz. H_3 model with Wavelet level 6 was selected which gives the maximum values with correct recognition rate 94.67 % and noise power of interfering sound 61 dB.

Table 5 The second experimental configuration

Parameter name	Value
Subject	Word classification
Feature extraction method	Discrete Wavelet Transform (DWT)
Wavelet level	*Variable parameter*
Wavelet function	Symlet 7 (sym7)
Network topology	*Variable parameter*
Node size in each layer	*Variable parameter*

Table 6 The second experimental results

Wavelet level	Network topology	Node size in each layer	$P_{noise,dB}$	Recognition rate (%)
1	H_1	{1000,18}	0.00	90.89
2	H_1	{1000,18}	33.00	93.56
1	H_2	{1000,1000,18}	0.00	90.44
2	H_2	{1000,1000,18}	29.75	93.33
3	H_2	{1000,1000,18}	39.75	94.00
4	H_2	{1000,1000,18}	48.50	94.67
1	H_3	{1000,4000,1000,18}	0.00	90.22
2	H_3	{1000,4000,1000,18}	28.25	92.22
3	H_3	{1000,4000,1000,18}	38.25	94.89
4	H_3	{1000,4000,1000,18}	52.25	94.44
5	H_3	{1000,4000,1000,18}	54.00	94.00
6	H_3	{1000,4000,1000,18}	**61.00**	**94.67**
7	H_3	{1000,4000,1000,18}	**60.00**	**95.33**
8	H_3	{1000,4000,1000,18}	**62.25**	**95.78**

Finally, the last experiment was designed to verify the hypothesis which Wavelet Transform feature extraction is suitable for the voice recognition application instead of STFT as shown in Tables 7 and 8. It is apparent seen that the correct recognition rates and noise power of interfering sounds in DWT achieved to increase high recognition rates than of STFT by reason of DWT theoretically employs multi-resolution [10] lead to offers the main characteristics or identity of voice at low frequency boundary which depends on Wavelet function and length of input signal.

Table 7 The third experimental configuration

Parameter name	Value
Subject	Word classification
Feature extraction method	*Variable parameter*
Wavelet level	6
Wavelet function	Symlet 7 (sym7)
STFT windows	Hamming
STFT time slot	1 ms
STFT frequency separation	8
Network topology	3rd-order All-features-connecting topology (H_3)
Node size in each layer	{1000,4000,1000,18}

Table 8 The third experimental results

Feature extraction method	$P_{noise,dB}$	Recognition rate (%)
Discrete Wavelet Transform (DWT)	61.00	94.67
Short-time Fourier Transform (STFT)	0.00	88.67

5 Discussions

The summaries of the optimized parameters with both of word and gender classification were described in Table 9. It can be seen that the proposed voice recognition with the optimized parameters offered high correct recognition rate and noise power were 96.22 % and 65.5 dB which sufficient for word classification. Moreover, the proposed voice recognition with the optimized parameters offered the correct recognition rate and noise power were 99.8 % and 72.25 dB which acceptable for gender classification.

The proposed voice recognition performance was established in term of the boundary of hit rate, false alarm and miss rate [8] with gender classification in order to compare with other models, i.e., simple sound database named Pitch-Cluster-Maps (PCMs). The performance of PCMs models established in term of Detection Error Tradeoff (DET) [4] curves with gender classification, in other words, it can be defines as upper and lower boundary both of false alarm rate and miss rate. The best performance of hit rate requires set of predicted data which approach to 100 % on true positive rate. In contrast, the best performance of false alarm and miss rate requires set of predicted data to approach on the false positive rate and false negative rate being closely equal to 0 and 0 %, respectively. Therefore, lower boundary of hit rate, upper boundary of false alarm rate and upper boundary of miss rate were important for performance evaluation. The proposed voice recognition performance was shown in Table 10.

On the one hand, subject to male classification, PCMs offered miss probability range from 2 to 12 % and false alarm probability range from 1 to 10 %, in other words, PCMs offered the upper boundary of miss and false alarm rate were 12 % and 10 %, respectively. Likewise, PCMs offered miss probability range from 2 to 20 % and false alarm probability range from 1 to 20 % for subject of female classification, in other words, PCMs offered the both of upper boundary of miss and

Table 9 The parameter optimization results

Parameter name	Subject	
	Word classification	Gender classification
Feature extraction method	Discrete Wavelet Transform (DWT)	Discrete Wavelet Transform (DWT)
Wavelet level	6	6
Wavelet function	Daubechies 15 (db15)	Daubechies 15 (db15)
Network topology	3rd-order All-features-connecting topology (H_3)	3rd-order All-features-connecting topology (H_3)
Node size in each layer	{1000,4000,1000,18}	{1000,4000,1000,2}
$P_{noise,dB}$	65.50	72.25
Recognition rate (%)	96.22	99.80

Table 10 Performance evaluation

Gender	Lower boundary	Upper boundary	
	Hit rate (%)	False alarm rate (%)	Miss rate (%)
Male	99.63	2.78	0.37
Female	97.22	0.37	2.78

false alarm rate were 20 %, which is appropriate for the female speaker identification from several words utterance.

On the other hand, proposed voice recognition with male classification offered upper boundary of miss and false alarm rate were 0.37 % and 2.78 %, respectively. With the female subjects, proposed voice recognition with male classification offered upper boundary of miss and false alarm rate were 2.78 % and 0.37 %, respectively, which is reduce the miss and false alarm rates leads to increases accuracy both of male and female classification with sufficient for work under normal conversation background noises conditions. However, the accuracy of gender classification is decreased since speech phase shift occurred. The variations of feature sets are further required for training the proposed voice recognition in order to implement large scale of word classification.

6 Conclusions

This paper presented an alternative voice recognition using combination of Artificial Neural Network and Multilevel of Discrete Wavelet Transform. The experimental results proved Wavelet Transform was achieved to increases high recognition rates up to 95 % instead of Short-time Fourier Transform feature extractions at noises up to 65 dB as in normal conversation background noises. The performance evaluation was demonstrated in terms of correct recognition rate, maximum noise power of interfering sounds, hit rate, false alarm rate and miss rate. The proposed method offers a potential alternative to intelligence voice recognition system in speech analysis-synthesis and recognition applications.

References

1. Furui, S.: 50 years of progress in speech and speaker recognition. In: SPECOM2005, pp. 1–9, Patras, Greece (2005)
2. Matsui, T., Tanabe, K.: Comparative study of speaker identification methods: dPLRM, SVM and GMM. In: IEICE Transactions on Information and System, vol. E89–D, no.3 (2006)
3. Matsui, T., Furui, S.: Comparison of text-independent speaker recognition methods using VQ-distortion and discrete/continuous HMMs. In: Acoustics, Speech, and Signal Processing, ICASSP, vol. 92. IEEE, San Francisco (1992)

4. Sasaki, Y., et al.: Pitch-cluster-map based daily sound recognition for mobile robot audition. J. Robot. Mechatron. **22**(3) (2010)
5. Zenteno, E., Sotomayor, M.: Robust voice activity detection algorithm using spectrum estimation and dynamic thresholding. IEEE Latin-American Communications, LATINCOM09 (2009)
6. Patil, S.P., Gowdy, J.N.: Exploiting the baseband phase structure of the voiced speech for speech enhancement. In: Acoustics, Speech and Signal Processing, ICASSP2014. IEEE, Florence, Italy (2014)
7. Krawczyk, M., et al.: Phase-sensitive real-time capable speech enhancement under voiced-unvoiced uncertainty. In: Signal Processing Conference, EUSIPCO2013, IEEE, Marrakech, Morocco (2013)
8. Bishop, C.M.: Pattern Recognition and Machine Learning (Information Science and Statistics). Springer, New York (2006). ISBN 978-0-387-31073-2
9. Møller, M.F.: A scaled conjugate gradient algorithm for fast supervised learning. Neural Netw. **6**(4), 525–533 (1993)
10. Mallat, S.G.: A theory for multiresolution signal decomposition: the wavelet representation. IEEE Pattern Anal. Mach. Intell. **11**(7), 674–693 (1989)
11. Bengio, Y., Lamblin, P., Popovici, D., Larochelle, H.: Greedy layer-wise training of deep networks. In: Advances in Neural Information Processing Systems (2007)
12. Bengio, Y.: Learning deep architectures for AI. Found. Trends Mach. Learn. **2**(1), 1–127 (2009)
13. Handbook of the International Phonetic Association: A Guide to the Use of the International Phonetic Alphabet. Cambridge University Press (1999). ISBN-0521637511

Parallel Dictionary Learning for Multimodal Voice Conversion Using Matrix Factorization

Ryo Aihara, Kenta Masaka, Tetsuya Takiguchi and Yasuo Ariki

Abstract Parallel dictionary learning for multimodal voice conversion is proposed in this paper. Because of noise robustness of visual features, multimodal feature has been attracted in the field of speech processing, and we have proposed multimodal VC using Non-negative Matrix Factorization (NMF). Experimental results showed that our conventional multimodal VC can effectively converted in a noisy environment, however, the difference of conversion quality between audio input VC and multimodal VC is not so large in a clean environment. We assume this is because our exemplar dictionary is over-complete. Moreover, because of non-negativity constraint for visual features, our conventional multimodal NMF-based VC cannot factorize visual features effectively. In order to enhance the conversion quality of our NMF-based multimodal VC, we propose parallel dictionary learning. Non-negative constraint for visual features is removed so that we can handle visual features which include negative values. Experimental results showed that our proposed method effectively converted multimodal features in a clean environment.

1 Introduction

Visual features have been attracted wide interest in the field of speech signal processing. One of the most famous examples is lip reading. Lip reading is a technique of understanding speech by visually interpreting the movements of the lips, face and tongue when the spoken sounds cannot be heard. For example, for people with hear-

R. Aihara (✉) · K. Masaka · T. Takiguchi · Y. Ariki
Graduate School of System Informatics, Kobe University, 1-1, Rokkodai, Nada, Kobe, Japan
e-mail: aihara@me.cs.scitec.kobe-u.ac.jp

K. Masaka
e-mail: makka@me.cs.scitec.kobe-u.ac.jp

T. Takiguchi
e-mail: takigu@kobe-u.ac.jp

Y. Ariki
e-mail: ariki@kobe-u.ac.jp

R. Lee (ed.), *Computer and Information Science*,
Studies in Computational Intelligence 656, DOI 10.1007/978-3-319-40171-3_3

ing problems, lip reading is one communication skill that can help them communicate better. McGurk et al. [17] reported that we perceive a phoneme not only from auditory information from the voice but also from visual information from the lips or from facial movements. Moreover, it is reported that we try to catch the movement of lips in a noisy environment and we misunderstand the utterance when the movements of the lips and the voice are not synchronized.

Visual information is used as noise robust features and audio-visual speech recognition has been studied for robust speech recognition under noisy environments [20, 33]. In audio-visual speech recognition, there are mainly three integration methods: early integration [21], which connects the audio feature vector with the visual feature vector; late integration [33], which weights the likelihood of the result obtained by a separate process for audio and visual signals, and synthetic integration [30], which calculates the product of output probability in each state and so on.

In [16], we have proposed an exemplar-based multimodal Voice Conversion (VC) method. In the task of automatic speech recognition (ASR), one problem is that the recognition performance remarkably decreases under noisy environments, and it becomes a significant problem seeking to develop a practical use of ASR. The same problem occurs in VC, which can modify nonlinguistic information, such as voice characteristics, while keeping linguistic information unchanged. The noise in the input signal is not only output with the converted signal, but may also degrade the conversion performance itself due to unexpected mapping of source features. In [16], the experimental result proved the effectiveness of visual features in the task of VC in noisy environments.

In our exemplar-based VC approach, Non-negative matrix factorization (NMF) [15], which is based on the idea of sparse representations, is a well-known approach for source separation and speech enhancement [25, 34]. In these approaches, the observed signal is represented by a linear combination of a small number of atoms, such as the exemplar and basis of NMF. In some approaches for source separation, the atoms are grouped for each source, and the mixed signals are expressed with a sparse representation of these atoms. By using only the weights of the atoms related to the target signal, the target signal can be reconstructed.

Our exemplar-based multimodal VC has major two problems. First, because our method is exemplar-based, the dictionary is over-compete and contains a large number of bases. Because lip movements closely resemble each other compared to speech spectra, lip images may be decomposed into a large number of bases, which can lead to a degradation of the converted sound. Second, because of non-negativity constraint in NMF, we added a constant value to visual features and this scheme degraded the conversion performance.

In order to tackle these problems, we propose parallel dictionary learning using matrix factorization. A compact parallel dictionary between the source and the target speakers are estimated using matrix factorization with parallel constraint. The non-negativity constraint for visual features is removed, where it enables to conduct the decomposition of the features which includes negative values. Our experimental results showed that our proposed method effectively converted multimodal features in a clean environment.

The rest of this paper is organized as follows: In Sect. 2, related works are described. In Sect. 3, our conventional exemplar-based method is described. In Sect. 4, our proposed method is described. In Sect. 5, the experimental data are evaluated, and the final section is devoted to our conclusions.

2 Related Works

VC is a technique for converting specific information to speech while maintaining the other information within the utterance. One of the most popular VC applications is speaker conversion [26] where a source speaker's voice individuality is changed to that of a specified target speaker such that the input utterance sounds as though the specified target speaker has spoken the utterance.

Other studies have examined several tasks that use VC. Emotion conversion is a technique that changes emotional information in input speech while maintaining linguistic information and speaker individuality [2, 32]. VC has also been adopted as assistive technology that reconstructs a speaker's individuality in electrolaryngeal speech [19], disordered speech [3] or speech recorded by non-audible murmur microphones [18]. Recently, VC has been used for ASR and speaker adaptation in text-to-speech (TTS) systems [12].

Statistical approaches to VC are the most widely studied [1, 26, 31]. Among these approaches, a GMM-based mapping approach [26] is the most common. In this approach, the conversion function is interpreted as the expectation value of the target spectral envelope, and the conversion parameters are evaluated using minimum mean-square error (MMSE) on a parallel training set. A number of improvements to this approach have been proposed. Toda et al. [29] introduced dynamic features and the global variance (GV) of the converted spectra over a time sequence. Helander et al. [10] proposed transforms based on partial least squares (PLS) in order to prevent the over-fitting problem associated with standard multivariate regression. In addition, other approaches that make use of GMM adaptation techniques [14] or eigen-voice GMM (EV-GMM) [22, 28] do not require parallel data.

We have proposed an exemplar-based VC method, which differs from conventional GMM-based VC. Exemplar-based VC using NMF has been proposed previously [27]. The advantage of NMF-based method can be summarized in two points. First, we assume that our NMF approach is advantageous in that it results in a more natural-sounding converted voice compared to conventional statistical VC. The natural sounding converted voice in NMF-based VC has been confirmed [5]. Wu et al. [35] applied a spectrum compression factor to NMF-based VC to improve conversion quality. Second, our NMF-based VC method is noise robust [4]. The noise exemplars, which are extracted from the before- and after-utterance sections in the observed signal, are used as the noise dictionary, and the VC process is combined with NMF-based noise reduction. NMF-based many-to-many VC has also been proposed [6].

NMF [15] is one of the most popular sparse representation methods. The goal is to estimate the basis matrix **W** and its weight matrix **H** from the input observation **V** such that:

$$\mathbf{V} \approx \mathbf{W}\mathbf{H}. \tag{1}$$

In this paper, we refer to **W** as the "dictionary" and **H** as "activity". NMF has been applied to hyperspectral imaging [7], topic modeling [11], and the analysis of brain data [8].

The NMF-based method can be classified into two approaches: the dictionary-learning approach and exemplar-based approach. In the dictionary-learning approach, the dictionary and the activity are estimated simultaneously. This approach has been widely applied in the field of audio signal processing: for example single channel speech separation [25, 34] and music transcription [24]. By estimating the dictionary from the training data, reconstruction errors between **V** and **WH** tend to be small. On the other hand, in the exemplar-based approach, only the activity becomes sparse because the dictionary is determined using exemplars and the activity is estimated using NMF. In the field of audio signal processing, Gemmeke et al. [9] proposed noise-robust automatic speech recognition using exemplar-based NMF. The disadvantage of this approach is the reconstruction error between **V** and **WH**, which becomes larger than the dictionary learning approach.

Because the conventional NMF-based approach employs exemplar-based NMF, the reconstruction error tends to be large, and we assume it results in "muffling effect". In order to enhance the conversion quality of NMF-based VC, we propose parallel dictionary learning for the multimodal VC method. Parallel dictionary learning using a matrix factorization method enables to obtain a compact dictionary of the source and target speakers and it enables to obtain better conversion quality compared to the conventional exemplar-based approach.

3 Exemplar-Based Multimodal Voice Conversion

3.1 Dictionary Construction

Figure 1 shows the process for constructing a parallel dictionary. To construct a parallel dictionary, some pairs of parallel utterances are required, with each pair consisting of the same text. The source audio dictionary $\mathbf{W}^{sa}$ consists of source speaker's audio features, source visual dictionary $\mathbf{W}^{sv}$ consists of source speaker's visual features, and the target dictionary $\mathbf{W}^{ta}$ consists of only target speaker's audio features. For audio features, a simple magnitude spectrum calculated by short-time Fourier transform (STFT) is extracted from clean parallel utterances. Mel-frequency cepstral coefficients (MFCCs) are calculated from the STRAIGHT spectrum to obtain alignment information in DTW. For visual features, DCT of lip motion images of

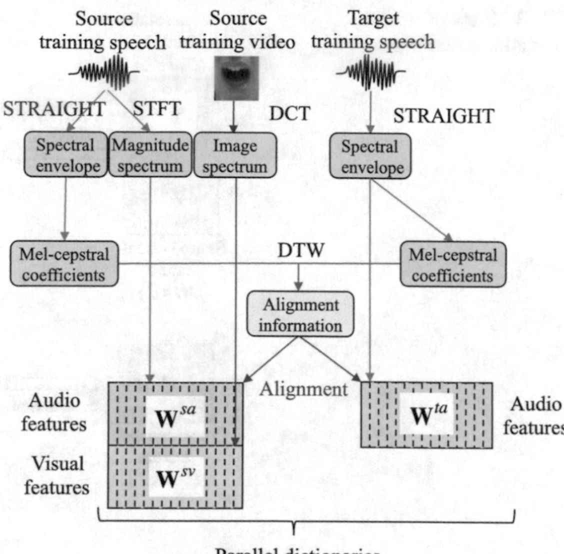

Fig. 1 Multimodal dictionary construction

the source speakers utterance is used. We have adopted 2D-DCT for lip images and perform a zigzag scan to obtain the 1D-DCT coefficient vector. Note that DCT coefficients contain negative data; therefore, we added a constant value to satisfy the non-negativity constraint of NMF such that the scale of the frame data is not changed. The visual features extracted from images taken by commonly-used cameras are interpolated by spline interpolation to fill the sampling rate gap between audio features. Aligned audio and visual features of the source speaker are joined and used as a source feature. The source and target dictionaries are constructed by lining up each of the features extracted from parallel utterances.

3.2 Conversion Flow

Figure 2 shows the basic approach of the conventional exemplar-based VC using NMF. Here D_{sa}, D_{sv}, D_{ta}, L, and J represent the number of dimensions of source audio features, the dimensions of source visual features, the dimensions of target audio features, the frames of the dictionary and the basis of the dictionary, respectively.

An input source audio matrix $\mathbf{V}^{sa}$ and an input source visual matrix $\mathbf{V}^{sv}$ are decomposed into a linear combination of bases from the source audio dictionary $\mathbf{W}^{sa}$ and the source visual $\mathbf{W}^{sv}$ using NMF. The dictionaries are determined with exemplars, and only the source activity $\mathbf{H}^{s}$ is estimated from the following cost function:

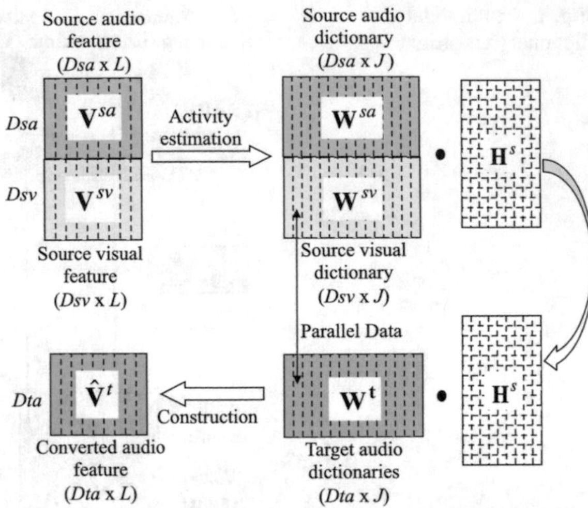

Fig. 2 Flow of exemplar-based VC

$$d_{KL}(\mathbf{V}^{sa}, \mathbf{W}^{sa}\mathbf{H}^{s}) + \psi d_{KL}(\mathbf{V}^{sv}, \mathbf{W}^{sv}\mathbf{H}^{s}) + \lambda||\mathbf{H}^{s}||_1 \quad s.t.\ \mathbf{H}^{s} \geq 0 \tag{2}$$

In (2), the first term is the Kullback-Leibler (KL) divergence between $\mathbf{V}^{sa}$ and $\mathbf{W}^{sa}\mathbf{H}^{s}$, the second term is the KL divergence between $\mathbf{V}^{sv}$ and $\mathbf{W}^{sv}\mathbf{H}^{s}$, and the third is the sparsity constraint with the L1-norm regularization term that causes the activity matrix to be sparse. ψ and λ represent the weight of the visual feature and the sparsity constraint. In order to fulfill the non-negative constraints, positive values are added to the input visual feature $\mathbf{V}^{sv}$ and the source visual dictionary $\mathbf{W}^{sv}$.

This function is minimized by iteratively updating the following equation.

$$\mathbf{H}^{s} \leftarrow \mathbf{H}^{s}. * \mathbf{F}./\mathbf{G} \tag{3}$$

$$\mathbf{F} = (\mathbf{W}^{sa\mathsf{T}}(\mathbf{V}^{sa}./(\mathbf{W}^{sa}\mathbf{H}^{s}))) + (\psi\mathbf{W}^{sv\mathsf{T}}(\mathbf{V}^{sv}./(\mathbf{W}^{sv}\mathbf{H}^{s}))) \tag{4}$$

$$\mathbf{G} = \mathbf{W}^{sa\mathsf{T}}\mathbf{1}^{(D_{sa}\times J)} + \psi\mathbf{W}^{sv\mathsf{T}}\mathbf{1}^{(D_{sv}\times J)} + \lambda\mathbf{1}^{(J\times L)} \tag{5}$$

$.*, ./$ and $\mathbf{1}$ denote element-wise multiplication, division and all-one matrix, respectively. In this sense, the input spectra are represented by a linear combination of a small number of bases and the weights are estimated as activity.

This method assumes that when the source signal and the target signal (which are the same words but spoken by different speakers) are expressed with sparse representations of the source dictionary and the target dictionary, respectively, the obtained activity matrices are approximately equivalent. The estimated source activity $\mathbf{H}^{s}$ is multiplied to the target dictionary $\mathbf{W}^{ta}$ and the target spectra $\hat{\mathbf{V}}^{t}$ are constructed.

$$\hat{\mathbf{V}}^{t} = \mathbf{W}^{ta}\mathbf{H}^{s} \tag{6}$$

3.3 Problems

We assume our exemplar-based approach contains two major problems. First, the dictionaries of the source and target speakers are over-complete and it results in the reconstruction error between $\mathbf{V}^s$ and $\mathbf{W}^s\mathbf{H}^s$. The error may cause "muffling effect" VC. Second, because of non-negativity constraint, we added a constant value to DCT, which is extracted from lip images. This scheme also degraded the conversion performance of VC.

4 Parallel Dictionary Learning for Multimodal Voice Conversion

4.1 Dictionary Learning

In order to construct a compact dictionary, a parallel dictionary between the source and target speakers is estimated using parallel-constrained matrix factorization. Figure 3 shows the flow of parallel dictionary learning for multimodal VC. $\mathbf{V}^{sa} \in \mathbb{R}^{+}$, $\mathbf{V}^{sv} \in \mathbb{R}^{\pm}$, and $\mathbf{V}^{ta} \in \mathbb{R}^{+}$ denote input the source speaker's audio exemplar, the source speaker's visual exemplar, and the target speaker's audio exemplar, respectively. These features are aligned using DTW. $\mathbf{W}^{sa} \in \mathbb{R}^{+}$, $\mathbf{W}^{sv} \in \mathbb{R}^{\pm}$, $\mathbf{W}^{ta} \in \mathbb{R}^{+}$ denote the source speaker's audio dictionary, the source speaker's visual dictionary, and the target speaker's audio dictionary, respectively. $\mathbf{H}^s \in \mathbb{R}^{+}$ and $\mathbf{H}^t \in \mathbb{R}^{+}$ denote the source activity, and the target activity, respectively. D_{sa}, D_{sv}, and D_{ta} denote the

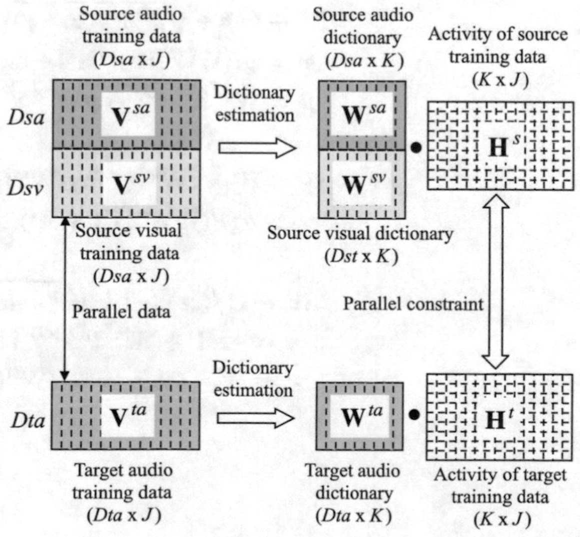

Fig. 3 Flow of parallel dictionary learning

number of dimension of the source audio feature, the source visual feature, and the target visual feature, respectively.

The objective function is represented as follows:

$$\begin{aligned} \min \quad & d_{KL}(\mathbf{V}^{sa}, \mathbf{W}^{sa}\mathbf{H}^{s}) + \frac{\psi}{2} d_F(\mathbf{V}^{sv}, \mathbf{W}^{sv}\mathbf{H}^{s}) \\ & + d_{KL}(\mathbf{V}^{ta}, \mathbf{W}^{ta}\mathbf{H}^{t}) \\ & + \frac{\varepsilon}{2}||\mathbf{H}^{s} - \mathbf{H}^{t}||_F + \lambda||\mathbf{H}^{s}||_1 + \lambda||\mathbf{H}^{t}||_1 \\ s.t. \quad & \mathbf{W}^{sa} \geq 0, \mathbf{H}^{s} \geq 0, \mathbf{W}^{ta} \geq 0, \mathbf{H}^{t} \geq 0 \end{aligned} \tag{7}$$

In (7), the first term is the KL divergence between $\mathbf{V}^{sa}$ and $\mathbf{W}^{sa}\mathbf{H}^{s}$, the second term is the Frobenius norm between $\mathbf{V}^{sv}$ and $\mathbf{W}^{sv}\mathbf{H}^{s}$, and the third therm is the KL divergence between $\mathbf{V}^{ta}$ and $\mathbf{W}^{ta}\mathbf{H}^{s}$. Because visual features include negative values, their divergence is restricted to the Frobenius norm. ψ denotes the weight for the visual feature. The forth term in (7) is the parallel constraint between $\mathbf{H}^{s}$ and $\mathbf{H}^{t}$, which is weighted by ε. The sparse constraints for activities are added as the fifth and sixth terms, which are weighted by λ.

This function is minimized by iteratively updating the following equations, which are derived from a majorization-minimization algorithm.

$$\begin{aligned} \mathbf{W}^{sa} &\leftarrow \mathbf{W}^{sa}. * ((\mathbf{V}^{sa}./(\mathbf{W}^{sa}\mathbf{H}^{s}))\mathbf{H}^{s^{\mathsf{T}}}) \\ &\quad ./(\mathbf{1}^{(D_{sa}\times L)}\mathbf{H}^{s^{\mathsf{T}}}) \end{aligned} \tag{8}$$

$$\mathbf{W}^{sv} \leftarrow (\mathbf{V}^{sv}\mathbf{H}^{s^{\mathsf{T}}})/(\mathbf{H}^{s}\mathbf{H}^{s^{\mathsf{T}}}) \tag{9}$$

$$\begin{aligned} \mathbf{W}^{ta} &\leftarrow \mathbf{W}^{ta}. * ((\mathbf{V}^{ta}./(\mathbf{W}^{ta}\mathbf{H}^{t}))\mathbf{H}^{t^{\mathsf{T}}}) \\ &\quad ./(\mathbf{1}^{(D^{ta}\times L)}\mathbf{H}^{t^{\mathsf{T}}}) \end{aligned} \tag{10}$$

$$\mathbf{H}^{s} \leftarrow (-\beta + \sqrt{\beta^2 + 4(\alpha. * \gamma)})./(2\alpha) \tag{11}$$

$$\alpha = \psi((\mathbf{W}\mathbf{W}^{+}\mathbf{H}^{s})./\mathbf{H}^{s}) + \varepsilon \tag{12}$$

$$\begin{aligned} \beta &= \mathbf{W}^{sa^{\mathsf{T}}}\mathbf{1}^{(D_{sa}\times J)} - \psi\mathbf{W}^{sv^{\mathsf{T}}}\mathbf{V}^{sv} \\ &\quad - \varepsilon\mathbf{H}^{t} + \lambda \end{aligned} \tag{13}$$

$$\begin{aligned} \gamma &= \mathbf{H}^{s}. * (\mathbf{W}^{sa^{\mathsf{T}}}(\mathbf{V}^{sa}./(\mathbf{W}^{sa}\mathbf{H}^{s}))) \\ &\quad + \psi((\mathbf{W}\mathbf{W}^{-}\mathbf{H}^{s}). * \mathbf{H}^{s}) \end{aligned} \tag{14}$$

$$\mathbf{H}^{t} \leftarrow (1/(2\varepsilon))(-\iota + \sqrt{\iota^2 + 4\varepsilon\kappa}) \tag{15}$$

$$\iota = -\varepsilon\mathbf{H}^{s} + \mathbf{W}^{ta^{\mathsf{T}}}\mathbf{1}^{(D_{ta}\times J)} + \lambda \tag{16}$$

$$\kappa = \mathbf{H}^{t}. * (\mathbf{W}^{ta^{\mathsf{T}}}(\mathbf{V}^{ta}./(\mathbf{W}^{ta}\mathbf{H}^{t}))) \tag{17}$$

In order to divide positive values and negative values, we define

$$\mathbf{WW}^{+} = \mathbf{W}^{+^{\top}}\mathbf{W}^{+}$$
$$\mathbf{WW}^{-} = \mathbf{W}^{-^{\top}}\mathbf{W}^{-}$$

where $\mathbf{X}^{+} = (|\mathbf{X}| + \mathbf{X})./2$ $\mathbf{X}^{-} = (|\mathbf{X}| - \mathbf{X})./2$.

4.2 Conversion

In conversion stage, the input source audio feature $\mathbf{V}^{sa}$ and the visual feature $\mathbf{V}^{sv}$ are decomposed into a linear combination of the bases from the estimated dictionaries.

The cost function is defined as follows:

$$\begin{aligned} &min\ d_{KL}(\mathbf{V}^{sa}, \mathbf{W}^{sa}\hat{\mathbf{H}}^{s}) \\ &\qquad + \frac{\psi}{2} d_{F}(\mathbf{V}^{sv}, \mathbf{W}^{sv}\hat{\mathbf{H}}^{s}) + \lambda||\hat{\mathbf{H}}^{s}||_{1} \\ &s.t\ \hat{\mathbf{H}}^{s} \geq 0 \end{aligned} \tag{18}$$

where $\hat{\mathbf{H}}^{s}$ denotes the activity of input source features. This function is minimized by iteratively updating the following equation.

$$\hat{\mathbf{H}}^{s} \leftarrow (-\hat{\beta} + \sqrt{\hat{\beta}^{2} + 4(\hat{\alpha}.*\gamma)})./(2\hat{\alpha}) \tag{19}$$

$$\hat{\alpha} = \psi((\mathbf{WW}^{+}\mathbf{H}^{s})./\mathbf{H}^{s}) \tag{20}$$

$$\hat{\beta} = \mathbf{W}^{sa^{\top}}\mathbf{1}^{(D^{sa}\times J)} - \psi\mathbf{W}^{sv^{\top}}\mathbf{V}^{sv} \tag{21}$$

The estimated source activity is multiplied to the target dictionary and the target spectra $\hat{\mathbf{V}}^{t}$ are constructed.

$$\hat{\mathbf{V}}^{t} = \mathbf{W}^{t}\hat{\mathbf{H}}^{s} \tag{22}$$

5 Experiment

5.1 Experimental Conditions

The proposed multimodal VC technique was evaluated by comparing it with an exemplar-based audio-input method [27] and exemplar-based multimodal method in a speaker-conversion task using clean speech data.

Table 1 Mel-cepstral distortion [dB] of each method

Source	Audio NMF	Multi NMF	Proposed
4.37	3.24	3.23	**3.17**

The source speaker was a Japanese male, and the target speaker was a Japanese female. The target female audio data were taken from the CENSREC-1-AV [23] database. We recorded the source male audio-visual data with the same text as the target female utterances. Table 1 shows the content of the audio data taken from the CENSREC-1-AV database. We used a video camera (HDR-CX590, SONY) and a pin microphone (ECM-66B, SONY) for recording. We recorded audio and visual data simultaneously in a clean environment. The camera was positioned 65 cm from the speaker and 130 cm from the floor.

Figures 4 and 5 show the recorded audio waves and lip images, respectively. We labeled the recorded audio data manually, and used the labeled data in a subsequent experiment. The sampling rate of the audio data in each database was 8 kHz, and the frame shift was 5 ms.

A total of 50 utterances of clean continuous digital speech were used to construct parallel dictionaries in our proposed method and the NMF-based methods. These utterances were also used to train the GMM in the GMM-based method. The other 15 words were used for testing.

In our proposed method and the NMF-based methods, a 257-dimensional magnitude spectrum was used for the source and noise dictionaries and a 512-dimensional STRAIGHT spectrum [13] was used for the target dictionary. The number of iterations used to estimate the activity was 300. In this paper, F0 information is converted

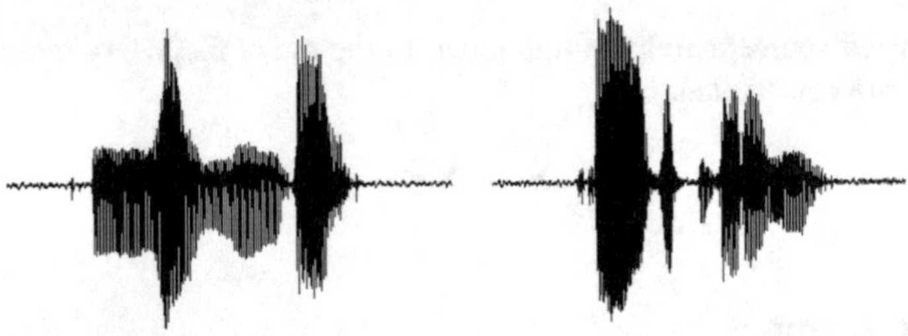

Fig. 4 Audio waves

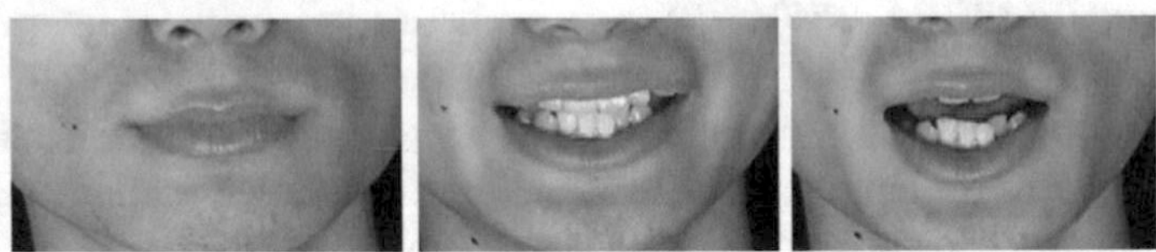

Fig. 5 Lip images

using a conventional linear regression based on the mean and standard deviation [29]. The other information, such as aperiodic components, is synthesized without any conversion.

The frame rate of the visual data was 30 fps. The lip image size was 130×80 pixels. For visual features, 50 DCT coefficients of the lip motion images of the source speaker's utterance were used. We introduced segment features for the DCT coefficient that consist of consecutive frames (two frames before and two frames after). Therefore, the total dimension of the visual feature was 250.

5.2 Experimental Results

Objective tests were carried out using Mel-cepstrum distortion (MelCD) [dB] as follows [29]:

$$MelCD = (10/\log 10)\sqrt{2\sum_{d}^{24}(mc_d^{conv} - mc_d^{tar})^2} \tag{23}$$

where mc_d^{conv} and mc_d^{tar} denote the dth dimension of the converted and target mel-cepstra.

First, we evaluated the effectiveness of visual feature in our proposed method. Figure 6 shows MelCD as a function of ψ in (7) and (18). As shown in this figure, $\psi = 0.1$ is an optimal value. When the visual feature is over weighted, it degrades the conversion quality.

Next, we evaluated the effectiveness of parallel constraint in our proposed method. Figure 7 shows MelCD as a function of ε in (7). As shown in this figure, we find $\varepsilon = 10$ is an a optimal value.

In Table 1, we compared our proposed method with the exemplar-based audio input NMF-based VC method and the exemplar-based multimodal VC method. Com-

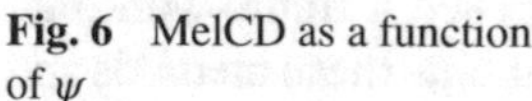
Fig. 6 MelCD as a function of ψ

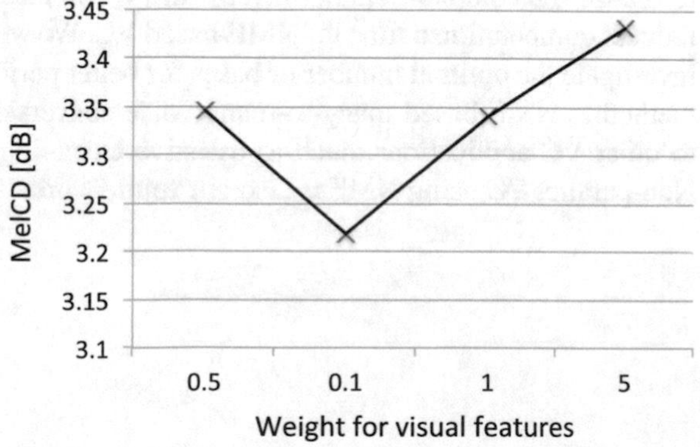

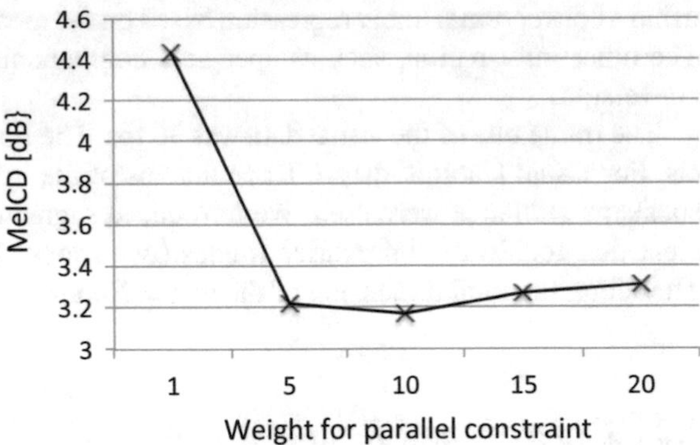

Fig. 7 MelCD as a function of ε

paring with audio NMF and multi NMF, the effectiveness of visual feature is quite small in the conventional exemplar-based method. However, our proposed method obtained the best result. We assume that our proposed method effectively enhanced NMF-based VC.

6 Conclusion

We proposed in this paper a parallel dictionary learning method for multimodal VC using matrix factorization. The major two problems for conventional multimodal VC are pointed. The first one is the over-complete parallel dictionary that causes "muffling effect" in converted voice. Second problems is non-negativity constraint for visual features. In order to tackle this problem, we proposed a parallel dictionary learning method, which estimate a compact dictionary of the source and target speakers. In the dictionary learning method, the non-negative constraint is removed and it enables to use visual features which include negative values.

In the future work, we will conduct subjective evaluation and further examine the effectiveness of our method. Wu et al. proposed a method for NMF-based VC to reduce the computational cost [35]. In [4], we also proposed a frame-work that reduces computational time for NMF-based VC. We will combine these methods and investigate the optimal number of bases for better performance. We will expand this method to NMF-based many-to-many voice conversion [6] and apply this method to other VC applications, such as assistive technology [5] or emotional VC [32]. Non-parallel VC using NMF is also our future work.

References

1. Abe, M., Nakamura, S., Shikano, K., Kuwabara, H.: Esophageal speech enhancement based on statistical voice conversion with Gaussian mixture models. In: ICASSP, pp. 655-658 (1988)
2. Aihara, R., Takashima, R., Takiguchi, T., Ariki, Y.: GMM-based emotional voice conversion using spectrum and prosody features. Am. J. Signal Process. **2**(5), 134–138 (2012)
3. Aihara, R., Takashima, R., Takiguchi, T., Ariki, Y.: Individuality-preserving voice conversion for articulation disorders based on non-negative matrix factorization. In: ICASSP, pp. 8037–8040 (2013)
4. Aihara, R., Takashima, R., Takiguchi, T., Ariki, Y.: Noise-robust voice conversion based on sparse spectral mapping using non-negative matrix factorization. IEICE Trans. Inf. Syst. **E97-D**(6), 1411–1418 (2014)
5. Aihara, R., Takashima, R., Takiguchi, T., Ariki, Y.: Voice conversion based on non-negative matrix factorization using phoneme-categorized dictionary. In: ICASSP, pp. 7944–7948 (2014)
6. Aihara, R., Takiguchi, T., Ariki, Y.: Multiple non-negative matrix factorization for many-to-many voice conversion. IEEE/ACM Trans. Audio Speech Lang. Process. (2016)
7. Berry, M.W., Browne, M., Langville, A.N., Pauca, V.P., Plemmons, R.J.: Algorithms and applications for approximate nonnegative matrix factorization. Comput. Stat. Data Anal. **52**(1), 155–173 (2007)
8. Cichocki, A., Zdnek, R., Phan, A.H., Amari, S.: Non-negative Matrix and Tensor Factorization. Wilkey (2009)
9. Gemmeke, J.F., Viratnen, T., Hurmalainen, A.: Exemplar-based sparse representations for noise robust automatic speech recognition. IEEE Trans. Audio Speech Lang. Process. **19**(7), 2067–2080 (2011)
10. Helander, E., Virtanen, T., Nurminen, J., Gabbouj, M.: Voice conversion using partial least squares regression. IEEE Trans. Audio Speech Lang. Process. **18**, 912–921 (2010)
11. Hofmann, T.: Probabilistic latent semantic indexing. In: Proceedings of the SIGIR, pp. 50–57 (1999)
12. Kain, A., Macon, M.W.: Spectral voice conversion for text-to-speech synthesis. In: ICASSP, pp. 285–288 (1998)
13. Kawahara, H.: STRAIGHT, exploitation of the other aspect of vocoder: perceptually isomorphic decomposition of speech sounds. Acoust. Sci. Technol. 349–353 (2006)
14. Lee, C.H., Wu, C.H.: MAP-based adaptation for speech conversion using adaptation data selection and non-parallel training. In: Interspeech, pp. 2254–2257 (2006)
15. Lee, D.D., Seung, H.S.: Algorithms for non-negative matrix factorization. Neural Inf. Process. Syst. 556–562 (2001)
16. Masaka, K., Aihara, R., Takiguchi, T., Ariki, Y.: Multimodal voice conversion based on non-negative matrix factorization. EURASIP J. Audio Speech Music Process. (2015)
17. McGurk, H., MacDonald, J.: Hearing lips and seeing voices. Nature **264**(5588), 746–748 (1976)
18. Nakamura, K., Toda, T., Saruwatari, H., Shikano, K.: Speaking aid system for total laryngectomees using voice conversion of body transmitted artificial speech. In: Interspeech, pp. 148–151 (2006)
19. Nakamura, K., Toda, T., Saruwatari, H., Shikano, K.: Speaking-aid systems using GMM-based voice conversion for electrolaryngeal speech. Speech Commun. **54**(1), 134–146 (2012)
20. Palecek, K., Chaloupka, J.: Audio-visual speech recognition in noisy audio environments. In: Proceedings of the International Conference on Telecommunications and Signal Processing, pp. 484–487 (2013)
21. Potamianos, G., Graf, H.P.: Discriminative training of HMM stream exponents for audio-visual speech recognition. In: ICASSP, pp. 3733–3736 (1998)
22. Saito, D., Yamamoto, K., Minematsu, N., Hirose, K.: One-to-many voice conversion based on tensor representation of speaker space. In: Interspeech, pp. 653–656 (2011)
23. Satoshi, T., Chiyomi, M.: Censrec-1-av an evaluation framework for multimodal speech recognition (japanese). Technical report, SLP-2010 (2010)

24. Sawada, H., Kameoka, H., Araki, S., Ueda, N.: Efficient algorithms for multichannel extensions of Itakura-Saito nonnegative matrix factorization. In: Proceedings of the ICASSP, pp. 261–264 (2012)
25. Schmidt, M.N., Olsson, R.K.: Single-channel speech separation using sparse non-negative matrix factorization. In: Interspeech (2006)
26. Stylianou, Y., Cappe, O., Moilines, E.: Continuous probabilistic transform for voice conversion. IEEE Trans Speech and Audio Processing **6**(2), 131–142 (1998)
27. Takashima, R., Takiguchi, T., Ariki, Y.: Exemplar-based voice conversion in noisy environment. In: SLT, pp. 313–317 (2012)
28. Toda, T., Ohtani, Y., Shikano, K.: Eigenvoice conversion based on Gaussian mixture model. In: Interspeech, pp. 2446–2449 (2006)
29. Toda, T., Black, A., Tokuda, K.: Voice conversion based on maximum likelihood estimation of spectral parameter trajectory. IEEE Trans. Audio Speech Lang. Process. **15**(8), 2222–2235 (2007)
30. Tomlinson, M.J., Russell, M.J., Brooke, N.M.: Integrating audio and visual information to provide highly robust speech recognition. In: ICASSP, pp. 821–824 (1996)
31. Valbret, H., Moulines, E., Tubach, J.P.: Voice transformation using PSOLA technique. Speech Commun. **11**, 175–187 (1992)
32. Veaux, C., Robet, X.: Intonation conversion from neutral to expressive speech. In: Interspeech, pp. 2765–2768 (2011)
33. Verma, A., Faruquie, T., Neti, C., Basu, S., Senior, A.: Late integration in audio-visual continuous speech recognition. In: ASRU (1999)
34. Virtanen, T.: Monaural sound source separation by non-negative matrix factorization with temporal continuity and sparseness criteria. IEEE Trans. Audio Speech Lang. Process. **15**(3), 1066–1074 (2007)
35. Wu, Z., Virtanen, T., Chng, E.S., Li, H.: Exemplar-based sparse representation with residual compensation for voice conversion. IEEE/ACM Trans. Audio Speech Lang. **22**, 1506–1521 (2014)

Unanticipated Context Awareness for Software Configuration Access Using the getenv API

Markus Raab

Abstract Configuration files, command-line arguments and environment variables are the dominant tools for local configuration management today. When accessing such program execution environments, however, most applications do not take context, e.g. the system they run on, into account. The aim of this paper is to integrate unmodified applications into a coherent and context-aware system by instrumenting the getenv API. We propose a global database stored in configuration files that includes specifications for contextual interpretations and a novel matching algorithm. In a case study we analyze a complete Debian operating system where every getenv API call is intercepted. We evaluate usage patterns of 16 real-world applications and systems and report on limitations of unforeseen context changes. The results show that getenv is used extensively for variability. The tool has acceptable overhead and improves context-awareness of many applications.

1 Introduction

The goal of context-oriented programming (COP) is to avoid the tedious, time-consuming and error-prone task of implementing context awareness manually, and instead adapt the application's behavior using the concept of layers [1, 12]. Each layer represents one dimension of the context relevant to the application. *Contextual values* [26] act as variables whose values depend on layers. A *program execution environment* consists of the environment variables and key/value pairs retrieved from configuration files. A program execution environment can be tightly integrated with contextual values [20]. *Context awareness* [5] is a property of software and refers to its ability to correctly adapt to the current context. Our aim is to make applications context-aware that previously were not.

M. Raab (✉)
Institute of Computer Languages, Vienna University of Technology, Vienna, Austria
e-mail: markus.raab@complang.tuwien.ac.at

R. Lee (ed.), *Computer and Information Science*,
Studies in Computational Intelligence 656, DOI 10.1007/978-3-319-40171-3_4

For example, an important context for a browser is the network it uses. In a different network, different proxy settings are required to successfully retrieve a web page. We want the browser to automatically adapt itself to the network actually present, i.e., make it context-aware in respect to the network.

Although COP eases the writing of new software, there remains a huge corpus of legacy software that cannot profit from context awareness. Our paper aims at intercepting the standard API `getenv` in a way that COP-techniques are applied to unmodified applications. We focus on `getenv` because we found that it is used extensively. Our interception technique, however, does not make any assumption on the API. We recommend to specify the values and the context of the program execution environments separately. This configuration specification contains placeholders, each representing a dimension of the context:

```
[/phone/call/vibration]
  type=boolean
  context=/phone/call/%inpocket%/vibration
```

In this example, `vibration` is a contextual value of type `boolean` and `%inpocket%` a placeholder to be substituted in contextual interpretations. Thus, the value of `vibration` changes whenever `inpocket` changes. E.g., when a context sensor measures body temperature only on one side of the gadget, it will change the value of `%inpocket%`. Thus, when the mobile phone is in the pocket, it will turn on vibration. When the mobile phone is lying on a table, it will turn off vibration to prevent falling down when someone calls. If needed, users can even specify further context. For example, some users dislike the context-dependent feature as described. Our approach inherently allows users to reconfigure every parameter in every context. To turn on vibration if the phone is *not* in the pocket, we configure our device differently:

```
/phone/call/inpocket/vibration = off
/phone/call/notinpocket/vibration = on
```

In this paper we analyze the popular `getenv()` API. The function `getenv()` is standardized by SVr4, POSIX.1-2001, 4.3BSD, C89, and C99. Because of this standardization and ease of use it is adopted virtually everywhere, even in core libraries such as `libc`. It allows developers to query the environment. Using standard `getenv` implementations developers have to act carefully: settings valid in the current context can differ from those received through `getenv`. To reduce the danger of assuming wrong context information we propose to use a context-aware implementation. We implement it in the whole system by intercepting every `getenv` API call. Our contributions are:

- We allow unmodified applications to use contextual values. In these standard applications the developers did not initially think of context awareness.
- We conduct an extensive case study and analyze 16 applications and systems.

These contributions are of practical relevance. While other approaches require code rewriting [3, 4], our approach is suitable for legacy applications, flexible and open for extensions. We tackle the research question: "How can we integrate unmodified applications into a coherent, context-aware system?"

The paper is structured as follows: In Sect. 2 we elaborate on the background. In Sect. 3 we explain our approach and in Sect. 4 we evaluate it. The validity of the evaluation is discussed in Sect. 5. After considering related work in Sect. 6 we conclude the paper in Sect. 7.

2 Background

Context-oriented programming (COP) enables us to naturally separate multi-dimensional concerns [5, 23, 25]. In some sense it extends object-oriented programming. Activation and deactivation of *layers* belong to its main concepts. Every layer represents a dimension of context that cuts across the system. All active layers together form the context the program currently is in.

The (de)activation of layers occur at any time during program execution. A currently active stack of layers determines the context the program or thread is in. COP allows us to specify programs with adaptable, dynamic behavior. Later approaches [27] go beyond object-oriented programming: they support program construction with layers only. Furthermore, later work considers software engineering perspectives [23] and modularity visions [13].

Tanter suggested a lightweight subset of COP: *Contextual values*. They are easier to understand because they "boil down to a trivial generalization of the idea of thread-local values" [26]. They are variables whose values depend on the current context. Contextual values originate from COP and naturally work along with the concepts of dynamic scoping and layers.

For newly written context-aware software, COP is a viable choice. For legacy software, however, rewriting seems unrealistic. So in this paper we introduce a new approach that does not require modifications of the application.

3 EnvElektra

In our approach, we want to intercept every call to the `getenv` API. Whenever an application calls the API, we want to invoke a context-aware implementation instead. EnvElektra, which is our research tool, contains such a `getenv()` implementation. The implementation contains a novel matching algorithm for context awareness. When EnvElektra is installed and activated on a system, the matching algorithm will be used for every call of `getenv()` done by any application.

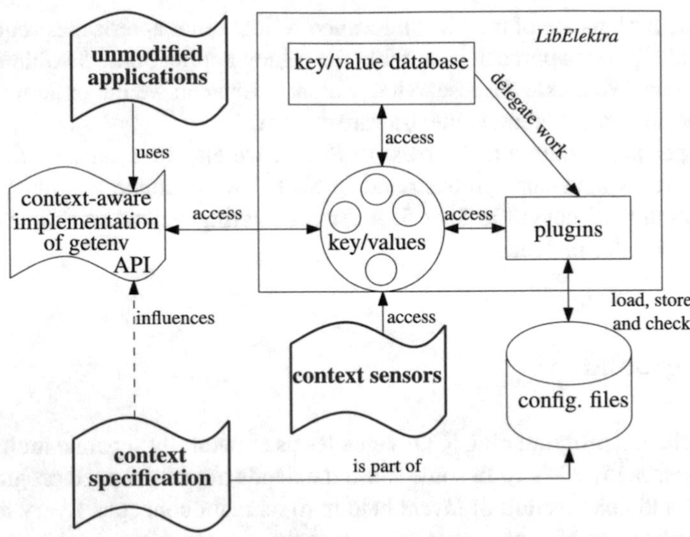

Fig. 1 Architecture of EnvElektra. The common data structure is a set of key/value pairs (*middle*). *Bold, blue boxes* need to be provided by users of EnvElektra

The basic idea of EnvElektra's `getenv()` implementation is as follows: First, it ensures that the data structure is up-to-date. Second, the matching algorithm calculates a new key for the parameter of `getenv()` using the context specification. Third, this key is searched in the data structure. With the found key, we recursively descend until every relevant context is considered.

The library *LibElektra* [20] (shown in Fig. 1) maps the *program execution environments* (e.g., command-line arguments and configuration files) to the in-memory key/value pairs. LibElektra includes start-up code that initializes all key/value pairs from a key database. The key database is modular via plugins [17]. The plugins allow us to use different syntax for configuration files.

Figure 1 also depicts the EnvElektra architecture. The system with EnvElektra has to provide three artifacts (bold, blue boxes): (1) unmodified applications that become context-aware, (2) context specifications, and (3) context sensors for out-of-process layer (de)activation. In the remainder of this chapter we will explain the user-provided artifacts and the matching algorithm. Finally, we will give a full example demonstrating how the system works interconnected.

3.1 Context Sensors

An essential issue to enjoy global, context-aware configuration access without modifying the application is an out-of-process layer (de)activation. We will show why such context sensors require us to use a database.

The original function getenv() retrieves values from the environment. Internally, it uses the data structure char** environ. By design, environ is copied into every process and will not receive any external changes afterwards. Thus, environ cannot consider out-of-process changes and cannot be used in EnvElektra.

We prefer to use configuration files that are read by the application itself. Then security is correctly handled by the operating system. In EnvElektra the administrator decides which configuration files are used, possibly with different syntax for each file [19]. EnvElektra makes sure that all applications have the same global view of the system's configuration files leading to a consistently configured system. This way values returned by getenv() will not be different from values retrieved from configuration files. The configuration files are viewed as a key/value database suitable for getenv lookups.

Context sensors observe the system and change the database when they detect context changes. They are responsible to modify the layers accordingly. Context sensors write their layer information into /env/layer. The key /env/layer is part of the database and resides within one of the configuration files. The use of files enables out-of-process communication between context sensor and applications. Thus, context changes can have an immediate effect on applications.
We identified two different kinds of context sensors to be used with our approach:

Information within the Database: Quite often, the necessary value is already present in the database. For example, in Linux many syscalls and the /sys-file system already provide much information. Using plugins, these sources are easily embedded within the database. Then we only need a symbolic link from /env/layer to the correct key. For example, if /env/layer/nodename points to /syscall/uname/ nodename, then %nodename% will resolve to the nodename as returned by the uname system call. In EnvElektra we mount plugins into any part of the hierarchy [17].

Context Sensor Daemons: In other cases, we implement a daemon, i.e. an active process, that updates /env/layer. Doing so, we can implement hysteresis, value transformations, and even complex feedback control systems. For example, to update %inpocket% a daemon measures the temperatures and modifies /env/layer/inpocket whenever we cross a threshold value. Changes in the database influence all processes across the whole system.

3.2 Context Specification

Up to now, we have established a database that contains key/value pairs to be used in a getenv() implementation. We have to make the database context-aware with the layer-information present in /env/layer, e.g.:

```
/env/layer/inpocket = notinpocket
```

Furthermore, we specify which key is used in which contextual interpretation:

```
[/phone/call/vibration]
  type=boolean
  context=/phone/call/%inpocket%/vibration
```

Now, when an API accesses `/phone/call/vibration`, the lookup layer will search for `/phone/call/%inpocket%/vibration`. Layer interpretations are stored in the database below the key `/env/layer`. In this case the correct contextual interpretation of `%inpocket%` is `notinpocket`. Using more than one placeholder creates several dimensions of variability. Late-binding is necessary so that unmodified software benefit from contextual features. EnvElektra needs to resolve its context awareness as late as possible, i.e., on `getenv()` calls.

For example, if a phone-call application executes `getenv("vibration")` it will look up `/phone/call/vibration`. Because of the `context` specification, we know we want the key `/phone/call/%inpocket%/vibration` instead. For the correct interpretation of `%inpocket%` we will lookup `/env/layer/inpocket` first. We get the value `notinpocket` for the layer `%inpocket%`. Thus, `getenv("vibration")` will return the value of `/phone/call/notinpocket/vibration`.

3.3 Matching Algorithm

The core of our approach is the contextual lookup within our alternative implementation of the `getenv` API. In EnvElektra `getenv()` provides the context-aware variability. The essence of EnvElektra's `getenv()` implementation is:

```
char* getenv(char* key) {
    if(needsReload(conf)) {
        reloadConfiguration(conf);
        reloadLayers(conf);
    }
    return contextLookup(conf, key);
}
```

Context is not static but dynamically changes over time. Our approach supports dynamic changes of context using `reloadLayers()` even though the original `getenv` implementation did not. The interception approach limits us to context-changes within `getenv()`: We cannot (de)activate layers at other places. Instead, we make sure that for every `contextLookup()` the correct context is used. The matching algorithm `contextLookup()` is recursively defined:

```
char* contextLookup (KeySet* cfg, char* key) {
    m = lookupBySpecification (cfg, key, "context");
    if (m) return contextLookup (cfg, fix(m));
    else return lookup (cfg, key);
}
```

The idea of the algorithm is: First, we look whether a context is specified for the key. If it is, `contextLookup` descends recursively after replacing all placeholders in the key. If it is not, a ordinary lookup will be used. The full implementation features namespaces, symbolic links and defaults [19].

3.4 Example

We present a full example that demonstrates recursion with several layers. Suppose a mobile phone is lying on the table in a building during a meeting. To simplify the example, we assign constant values to the layers:

```
/env/layer/inpocket = notinpocket
/env/layer/inbuilding = inbuilding
/env/layer/inmeeting = inmeeting
```

In a real system, a sensor will continuously update the values. So far, we already discussed the layer `inpocket`. The layer `inbuilding` represents a value from a location context. Layers such as `inmeeting` are called *virtual sensors* [1]. In this case the value of the layer is calculated by a sensor querying the person's schedule. The application running on the phone uses the following non-context-aware code:

```
char* use_vibration = getenv("vibration");
if (!strcmp(use_vibration, "on")) {/* activate vibration */}
```

We add context awareness with the following specification:

```
[/phone/call/vibration]
  type=boolean
  context=/phone/call/%inbuilding%/vibration
[/phone/call/inbuilding/vibration]
  type=boolean
  context=/phone/call/%inpocket%/%inmeeting%/vibration
[/phone/call/notinbuilding/vibration]
  type=boolean
  context=/phone/call/%handsfree%/vibration
```

Due to lack of space, we here specify only two of the six possible configurations:

```
/phone/call/inpocket/inmeeting/vibration = on
/phone/call/notinpocket/inmeeting/vibration = off
```

Suppose the mobile phone gets a call. By above `getenv` we request to lookup `/phone/call/vibration` to know whether vibration is turned on. In the first step, it will find the `context` and resolve `inbuilding`. In the next step, it will recursively search in the specification again, and find another `context` with `/phone/call/ %inpocket%/%inmeeting%/vibration`. Then the placeholders are again replaced with the respective values. Resolving this key, the algorithm will not find another matching specification. Thus, it returns the configuration value of not in pocket and in meeting, i.e., `/phone/call/notinpocket/inmeeting/vibration`. Because this configuration value is `off`, the phone will not vibrate.

4 Evaluation

Our methodological foundation is built on "theory of cases" [6, 7]. Other research should supplement our work with further case and user studies.

We chose 16 popular systems for evaluation (as discussed in threats to validity in Sect. 5). We will solely focus on existing applications and their integration into a coherent system.

The evaluation was conducted on different machines using Debian GNU/Linux Jessie 8.1 amd64. For the evaluation we globally intercept `getenv()` using `/etc/ld.so.preload`. By listing EnvElektra in `/etc/ld.so.preload` it will be loaded before any other library. Thus its symbols will be preferred. Because of this preference EnvElektra will be used for every `getenv()`-call.
In each of the following subsections, we will answer one of the questions:

RQ1: What are the usage patterns of `getenv()` in popular applications?
RQ2: For which applications can we actually exploit `getenv()` to be used for unanticipated context awareness? What are the fundamental limitations?
RQ3: What is the overhead that occurs in a system using EnvElektra?

4.1 RQ1: Usage Patterns

Only APIs that are actually called during runtime can be exploited for context awareness. To learn more about usage patterns, we count how often `getenv(key)` is executed.

application	lines of code	getenv all	getenv init	all unique	later unique	same
akonadi	37,214	10,357	8655	110	12	5126
chromium	18,032,183	6006	1803	1118	192	165
curl	249,380	19	8	12	8	4
eclipse	3,311,712	2790	2696	389	42	1495
evolution	672,789	4407	1488	1060	24	163
firefox	12,394,938	3371	2049	276	70	895
gimp	901,703	2551	1115	217	137	364
inkscape	479,849	722	457	160	51	166
libreoffice	5,482,215	3354	2891	258	59	1493
lynx	192,012	1931	961	27	27	923
man	142,183	2862	13	86	76	2
smplayer	76,170	212	164	71	8	53
wget	142,603	11	10	8	1	3
Mean	3,217,074	2969	1716	292	54	835
Median	479,849	2790	1115	160	42	166
Total	41,821,956	38,593	22,310	3792	707	10,852
KDE	*	*	9606	265	*	2634
GNOME	*	*	144	47	*	4
Debian	*	*	5317	430	*	286

* Any of the above applications can be started within the same session.

lines of code: Count lines of code with the tool `cloc`.
getenv all: Count all calls to getenv while using the application.
getenv init: Count all calls to getenv while starting the application.
all unique: From all getenv calls, how many different `keys` were used?
later unique: From getenv calls after initialization, how many different `keys` were used? For wget and curl the first download counts as initialization.
same: From the getenv calls during startup (during runtime an arbitrary high number could be acquired), what is the maximum number of queries with the same value for the parameter `key`?

To interpret the numbers correctly we have to know that the usage patterns vary widely even for the same application. For example, firefox started within GNOME requests 11 GNOME specific and 8 GTK specific environment variables (like `G_DEBUG`). If executed on a system with OpenGL enabled, 43 additional environment variables (like `__GL_EVENT_LOGLEVEL`) are used to determine OpenGL configurations. Additionally, the tested system requested three vendor (`NV`) specific variables. For KDE, `KDE_FULL_SESSION` was used as detection. Then 8 more KDE-specific and 15 more QT-specific environment variables were requested if started within KDE. Thus, the numbers depend on the desktop environment and hardware.

For better reproducibility, we freshly installed Debian Jessie KDE and GNOME variants, respectively. The only modification was the installation of EnvElektra. For example, on a daily used KDE with many installed applications, we measured 210.276 `getenv()` during startup, which is 21 more than with a freshly installed KDE. We see that the numbers also depend on the installed software.

The above 13 applications request an average of 2969 environment values (2790 median). Akonadi, configured to use IMAP, had the highest number of calls to `getenv`. The reason seems to be a potential misuse of a libc function which requested `LANGUAGE` 5126 times. During the KDE startup 27 % of all `getenv` calls were `LANGUAGE`. We conclude that excessive use can be unintentional.

From the numbers in the table we conclude that `getenv()` is used extensively in all examined applications. Applications often reread environment parameters during user interactions. This statement is true for both large applications and small helper tools. As expected, large feature-rich applications request much more environment variables. The ratios of requested and unique environment variables varies greatly: it is 14 % median, and in akonadi it is ~1 %. We see that applications tend to request the same variables often.

Our findings regarding **RQ1** are:

(1) We quantitatively show that `getenv()` is pervasive. We think that the usage patterns stem from a rather random use of `getenv()`: variability seems to be added ad-hoc whenever single developers needed it. Because `getenv()` has no noticeable performance implication and typically is not unit-tested, it is likely that quality assurance will not find unnecessary occurrences.

(2) Based on our observation, `getenv()` is used frequently after startup.

Implications: Developers seem to not optimize calls to `getenv()`. The resulting high number of `getenv()`-calls open up possibilities to influence the behavior of applications on context changes.

RQ2: Unanticipated Context Awareness

We already showed that the use of `getenv()` is pervasive, even after startup. Now, we want to find out whether changes in the context—and thus in the variables returned by `getenv()`—actually have an influence on the behavior.

We found that in help-, save- and open-dialogs different values returned by `getenv()` often influence the behavior of the application in a way easily visible to the user. These environment variables often have immediate and visible impact when changed dynamically. For example, gimp uses for every open dialog `G_FILENAME_ENCODING` and for every help dialog `GIMP2_HELP_URI`. On context changes, e.g. when we enter another network or mount a new file system, the software can automatically be adapted with EnvElektra.

Now, we investigate context awareness of proxy settings. A user changing the network with a different proxy should be able to continue browsing. `lynx` requests and correctly uses `http_proxy` for every single page. `curl` has the same behavior and reloads 7 additional environment variables every time. `wget` gives less control per download but still requests `http_proxy` for every page in recursive downloading mode. Firefox uses the proxy for most pages but pages in cache are displayed even when the proxy is unreachable. Chromium is the only browser not rereading `http_proxy`. Instead, it requests many internals such as `GOOGLE_API_KEY` during run-time. EnvElektra supports `http_proxy` well.

Our approach is very successful whenever an application executes other programs because during the startup of the programs the whole environment is always requested and used. Many programs use a pager or editor as external program. For example, `man` executes a pager for every displayed manpage.

For some applications it is possible to specify a configuration file using an environment variable. In EnvElektra configuration files can be mounted. Then they are a part of the database, which permits full configurability. For example, `less` executed within `man` uses the environment variable `LESSKEY`. In such cases our approach provides seamless context-aware configuration.

Some `getenv()` calls, however, do not have any user-visible impact. Instead, they seem to be left-overs. In LibreOffice, `WorkDirMustContainRemovableMedia` is obviously a workaround for a very specific problem. It is not documented and searching the web for it only reveals the use in the source code. Instead, `OOO_ENABLE_LOCALE_DATA_CHECKS` is an announced workaround. In GTK `GTK_TEST_TOUCHSCREEN` is requested extensively. According to the commit log it was explicitly introduced as a test feature.

Sometimes recurring `getenv` cannot be exploited to improve context awareness. For example, `LANGUAGE` is requested very often but does not influence the user-interface after startup. Here changes at runtime seem to have no impact. Such environment variables will only be context-aware during the start of an application.

A limitation of our approach is the impossibility to detect unwanted changes of environment variables. For example, the environment variable `CC` can change during compilation. Obviously, this easily leads to inconsistent compilation and linking. In EnvElektra the runtime-context-change feature can easily be (de)activated for process hierarchies, though.

Not a single crash occurred in our experiments regardless of which values we modified. This behavior is not entirely surprising: First, software should validate values returned from `getenv()`. Thus, wrong values from `getenv()` are rejected. Second, we did no systematic stress testing but only searched for useful changes.

> Our findings regarding **RQ2** are:
>
> (1) We show that many practical use cases exist where context changes are applied successfully at runtime.
>
> (2) Limitations include that some `getenv()` calls do not have visible impact and that context switches in rare cases lead to incorrect behavior.
>
> **Implications**: EnvElektra increases the context awareness for the evaluated applications. Specific functionality is even flawlessly context-aware.

RQ3: Overhead
Finally, we want to evaluate whether the overhead of EnvElektra is acceptable. The benchmarks were conducted on a hp® EliteBook 8570 w using the central processor unit Intel® Core™ i7-3740QM @ 2.70 GHz. Overhead is measured with valgrind by running the executable without and with EnvElektra.

The glibc `getenv()` implementation linearly searches through the whole environment. On the one hand, our implementation does not have this constraint. Its complexity is $O(\log(n))$ compared to $O(n)$ for `environ` iteration. We do not use unordered hash maps because we need lexically ordered iteration, e.g. to iterate over all layers and during `reloadConfiguration()`. On the other hand, the contextual lookup involves recursion. Depending on the specification EnvElektra needs additional nested lookups.

In a benchmark we compared 1,000,000 `getenv()` calls with the same number of EnvElektra's lookups. We did 11 measurements and report the median value. For a small number (30) of environment variables, standard `getenv()` implementations (0.03 s) clearly outperform EnvElektra's lookup (0.06 s). For 100 environment variables (which is a typical value) they perform equally well: 0.076 s for standard `getenv()` and 0.073 s for EnvElektra's lookup. For more than 100 environment variables, EnvElektra's lookup outperforms `getenv()`.

Regarding the overall overhead, we first report about the diversity of the applications. For the startup of gimp the overhead of 2.6 % is negligible. For the startup of firefox, however, the overhead is 6.5 %. The reason is that Firefox performs `exec()` 5 times during startup. Then EnvElektra needs to be initialized and needs to parse its configuration files again. For very small applications, e.g. `curl` and `wget`, the parsing strongly affects the runtime overhead. If they download empty files, the overhead even dominates. The overhead between different applications varies greatly.

Next, we were interested in the impact on a system which executes many processes each with trivial tasks. An extreme example happens to be the compilation of C software projects with gcc. Because gcc spawns 5 subprocesses for the compilation of every `.c` file, the overhead seems to get immense. Actually, the overhead of a trivial program's compilation, only containing `int main()`, is 90 %. The parsing of configuration files gets dominant. It is astonishing that the overhead of a compilation for a full project is only 14 %. For this benchmark we compiled EnvElektra from scratch. The absolute times are 2:23 min total when compiling with EnvElektra and 2:05 min total without EnvElektra as measured with the `time` utility. The compilation executed 6847 processes, did 30862 `getenv` calls, 6199 of which contained `CC`. Even though trivial process executions have large overhead, the overall performance only suffers little, even in extreme cases.

We further were very interested in any other occurrence with a similar number of many process executions. The booting of Debian executes 732 processes. The most often requested environment variable was `SANE_DEBUG_SANEI_SCSI` with 286 occurrences. In the script `startkde`, 227 binaries are executed. The executed number of processes in the case of compilation actually seems to represent an exception. We conclude that occurrences where processes are spawned excessively are rare.

Finally, we want to discuss the overhead of the reload feature. We chose the following setup: We installed the webserver lighttpd locally. EnvElektra was active throughout the whole experiment. To download 10 files with 1MB to 10MB size each we executed `curl -o"#1 http://localhost/test/[1-10]"`. Without reloading this execution resulted in 83,786,947 instructions. With reloading EnvElektra every millisecond, valgrind counted 91,569,790 instructions. The reloading caused the configuration to be fetched 91 times instead of 4 times. Because of an optimization within EnvElektra only `stat` is used on the configuration files without parsing them again. Thus, the overhead is only 9.3 %.

Different to the benchmark setup above we will now change the database once during program execution. Then EnvElektra will reread the respective configuration file. We have to take care that the changed value does not influence the control flow. For example, if we add the `no_proxy` variable, proxy setup is skipped and the performance even increases. Thus, we changed `COLUMNS`, which is requested for every download but does not influence the overhead more than unrelated parameters. When changing it during one of the ten requests the execution needed 95,248,722 instructions. We see that actual context changes have acceptable overhead of ~4 %.

Our findings regarding **RQ3** are:

(1) In applications that terminate very soon, e.g. only showing help text, the run-time overhead dominates. In practical use, however, EnvElektra only adds run-overhead from 2.6 to 14 % (in extreme but realistic cases).

(2) Dynamic reload has about 10 % overhead. On context changes the overhead increases again by about 4 % in a realistic http-proxy-transition.

Implications: EnvElektra's run-time overhead typically is low and thus acceptable. For frequent context changes, optimizations would be preferable.

5 Threats to Validity

As in all quantitative studies our concern is if the evaluated software is representative. In **RQ1** we address it by using a significant number of diverse open-source software in terms of functionality, development teams and programming languages. We did not consider context awareness already present in applications. Although interception also works for closed-source software, we did not study it because of the impossibility to cross-check with source code. Anyhow some of the software, including libreoffice, chromium and eclipse, has at least origins in closed-source development. Thus, the results can be valid for closed-source software, too. While we think that the software we inspected represents some characteristics of variability APIs, more general conclusions need further work.

In the methodology of **RQ2**, we need to interpret whether contextual awareness can be exploited. We avoid subjective judgements about context awareness during program start. One could also modify the environment with a wrapper script to achieve similar results. We prefer to examine dynamic context changes which are impossible with former approaches. To improve reproducibility and objectivity we only consider visible changes in the user interface.

We exclusively measure calls of `getenv` but do not consider the use of the `environ` pointer, the third parameter of `main`, and `/proc`. We cannot guarantee full coverage. Therefore our evaluation actually underestimates the full potential.

We added optional logging to count the number of `getenv`. Logging, however, influences a system deeply. On one system two start-processes failed when logging was activated. We did not find other occurrences that caused differences in behavior. Thus, we always rerun our tests without logging.

The benchmarks are conducted comparatively and consider only a single implementation of `getenv`. Therefore run-time measurements may not apply for other versions or OSs. Additionally, the benchmarks yield very different results depending on the size of the used configuration files and the respective parser. To level out this problem, we took care that our setup is realistic. We used 8 different configuration files and especially chose parsers which are known to be slow. We think that it is straight-forward to reproduce our benchmarks in a way that they perform even better than the numbers we reported.

Overall, while we cannot draw general conclusions for context-aware configuration access in the `getenv` API, we think that our study unveils some important insights, particularly for open source software.

6 Related Work

Riva et al. [22] acquired software-engineering-related knowledge from studying context-aware software. Different from our approach, they reverse-architected existing context-aware support systems. We preferred to study the behaviour of well-known software when introducing context awareness.

Context-aware middleware [8, 9] is a well-established research direction. EnvElektra could be seen as local context-aware middleware for configuration. EnvElektra scores in situations where legacy software needs to be deployed.

Using the correct context is a subtopic of avoiding configuration errors. Yin et al. [28] researched different types of configuration-parameter-related mistakes. They investigated value-environment mistakes which can be caused by wrong contextual interpretation. Which errors actually are induced by incorrect contextual interpretation, however, is still an open question.

A lot of work exists about how to extract program configuration constraints from source code [15, 21]. The authors argue that even though many constraints are extracted, sometimes additional external knowledge is needed. We think that context awareness is such a constraint.

Context-oriented programming (COP) already has an important role within software-engineering [1, 12, 23]. COP mainly aims at more comprehensible programs expressing more context awareness. Our approach tackles the problem in a different direction: We add context awareness without changing the program.

Previous work [18] describes context-awareness by using explicit layer activations. Other than our approach, these methods cannot be used for already existing applications.

Niu et al. [16] report on a web-based framework which uses indoor location, which is an important context sensor. Software product line engineering [2, 24] deals with the question how to construct products by combining features. Configuration specification languages [10, 11] rarely have support for context. An exception is the context oriented component model PCOM [14]. Unlike our approach, these approaches cannot be used for already existing applications.

Yuan et al. [29] provided a quantitative characteristic study for software logging. Similar to our study they revealed that their object of study is used in four large open-source applications pervasively. Different to our approach, they researched how logging statements were introduced and changed, while we show how APIs for variability are intercepted for more context awareness.

7 Conclusion

In this paper, we described a context-aware database using configuration files. A `getenv` implementation uses it for context-aware configuration access. Applications facilitating this API profit from context awareness. Our approach is unique because it allows applications to be context-aware without any modifications.

We saw that `getenv()` in most software provides excessive variability which is currently underutilized. This variability benefits from context awareness. The paper gives ideas for programmers how `getenv()` can be used with more efficacy. Sometimes software is even capable to dynamically adapt to context changes even though the authors did not anticipate this use. In a benchmark we found out that while in small synthetic benchmarks the overhead might be devastating, in practice it stays well with reasonable bounds.

Our results are:

- Presentation of an approach in which applications are more aware of their context
- A novel context-aware `getenv()` implementation downloadable from http://www.libelektra.org.
- Providing experimental validation by a case study of significant complexity.

Acknowledgments I would like to thank Franz Puntigam, Helmut Toplitzer, Christian Amsss, Nedko Tantilov and the anonymous reviewers for a detailed review of this paper. Many thanks especially to Natalie Kukuczka and Elisabeth Raab.

References

1. Baldauf, M., Dustdar, S., Rosenberg, F.: A survey on context-aware systems. Int. J. Ad Hoc Ubiquit. Comput. **2**(4), 263–277 (2007)
2. Berger, T., Lettner, D., Rubin, J., Grünbacher, P., Silva, A., Becker, M., Chechik, M., Czarnecki, K.: What is a feature?: a qualitative study of features in industrial software product lines. In: Proceedings of the 19th International Conference on Software Product Line, pp. 16–25. ACM (2015)
3. Bockisch, C., Kanthak, S., Haupt, M., Arnold, M., Mezini, M.: Efficient control flow quantification. In: ACM SIGPLAN Notices, vol. 41, pp. 125–138. ACM (2006)
4. Costanza, P., Hirschfeld, R., De Meuter, W.: Efficient layer activation for switching context-dependent behavior. In: Lightfoot, D., Szyperski, C. (eds.) Modular Programming Languages, Lecture Notes in Computer Science, vol. 4228, pp. 84–103. Springer (2006). http://dx.doi.org/10.1007/11860990_7
5. Dey, A.K., Abowd, G.D.: The what, who, where, when, why and how of context-awareness. In: CHI '00 Extended Abstracts on Human Factors in Computing Systems, CHI EA '00. ACM, NY (2000). ftp://ftp.cc.gatech.edu/pub/gvu/tr/1999/99-22.pdf
6. Easterbrook, S., Singer, J., Storey, M.A., Damian, D.: Selecting empirical methods for software engineering research. In: Shull, F., Singer, J., Sjøberg, D. (eds.) Guide to Advanced Empirical Software Engineering, pp. 285–311. Springer (2008). http://dx.doi.org/10.1007/978-1-84800-044-5_11
7. Eisenhardt, K.M., Graebner, M.E.: Theory building from cases: opportunities and challenges. Acad. Manag. J. **50**(1), 25–32 (2007)
8. Geihs, K., Barone, P., Eliassen, F., Floch, J., Fricke, R., Gjorven, E., Hallsteinsen, S., Horn, G., Khan, M.U., Mamelli, A., Papadopoulos, G.A., Paspallis, N., Reichle, R., Stav, E.: A comprehensive solution for application-level adaptation. Softw. Pract. Exp. **39**(4), 385–422 (2009). http://dx.doi.org/10.1002/spe.900
9. Gu, T., Pung, H.K., Zhang, D.Q.: A middleware for building context-aware mobile services. In: Vehicular Technology Conference, 2004. VTC 2004-Spring. 2004 IEEE 59th, vol. 5, pp. 2656–2660. IEEE (2004)
10. Günther, S., Cleenewerck, T., Jonckers, V.: Software variability: the design space of configuration languages. In: Proceedings of the 6th Workshop on Variability Modeling of Software-Intensive Systems, pp. 157–164. ACM (2012)
11. Hewson, J.A., Anderson, P., Gordon, A.D.: A declarative approach to automated configuration. LISA **12**, 51–66 (2012)
12. Jong-yi, H., Eui-ho, S., Sung-Jin, K.: Context-aware systems: a literature review and classification. Expert Syst. Appl. **36**(4), 8509–8522 (2009). http://dx.doi.org/10.1016/j.eswa.2008.10.071
13. Kamina, T., Aotani, T., Masuhara, H., Tamai, T.: Context-oriented software engineering: a modularity vision. In: Proceedings of the 13th International Conference on Modularity. MODULARITY '14, pp. 85–98. ACM, New York, NY, USA (2014)
14. Magableh, B., Barrett, S.: Primitive component architecture description language. In: 2010 The 7th International Conference on Informatics and Systems (INFOS), pp. 1–7 (2010)
15. Nadi, S., Berger, T., Kästner, C., Czarnecki, K.: Mining configuration constraints: Static analyses and empirical results. In: Proceedings of the 36th International Conference on Software Engineering, ICSE 2014, pp. 140–151. ACM, New York, NY, USA (2014). doi:10.1145/2568225.2568283
16. Niu, L., Saiki, S., Matsumoto, S., Nakamura, M.: Wif4inl: Web-based integration framework for indoor location. Int. J. Pervasive Comput. Commun. (2016)
17. Raab, M.: A modular approach to configuration storage. Master's thesis, Vienna University of Technology (2010)
18. Raab, M.: Global and thread-local activation of contextual program execution environments. In: Proceedings of the IEEE 18th International Symposium on Real-Time Distributed Computing Workshops (ISORCW/SEUS), pp. 34–41 (2015). doi:10.1109/ISORCW.2015.52

19. Raab, M.: Sharing software configuration via specified links and transformation rules. In: Technical report from KPS 2015. Vienna University of Technology, Complang Group, vol. 18 (2015)
20. Raab, M., Puntigam, F.: Program execution environments as contextual values. In: Proceedings of 6th International Workshop on Context-Oriented Programming, pp. 8:1–8:6. ACM, NY, USA (2014). http://dx.doi.org/10.1145/2637066.2637074
21. Rabkin, A., Katz, R.: Static extraction of program configuration options. In: 2011 33rd International Conference on Software Engineering (ICSE), pp. 131–140. IEEE (2011)
22. Riva, O., di Flora, C., Russo, S., Raatikainen, K.: Unearthing design patterns to support context-awareness. In: Fourth Annual IEEE International Conference on Pervasive Computing and Communications Workshops, 2006. PerCom Workshops 2006, pp. 5–387 (2006). http://dx.doi.org/10.1109/PERCOMW.2006.138
23. Salvaneschi, G., Ghezzi, C., Pradella, M.: Context-oriented programming: A software engineering perspective. J. Syst. Softw. **85**(8), 1801–1817 (2012). http://dx.doi.org/10.1016/j.jss.2012.03.024
24. Schaefer, I., Hähnle, R.: Formal methods in software product line engineering. IEEE Comput. **44**(2), 82–85 (2011)
25. Schippers, H., Molderez, T., Janssens, D.: A graph-based operational semantics for context-oriented programming. In: Proceedings of the 2nd International Workshop on Context-Oriented Programming, COP '10. ACM, NY, USA (2010). doi:10.1145/1930021.1930027
26. Tanter, E.: Contextual values. In: Proceedings of the 2008 Symposium on dynamic languages, DLS '08, pp. 3:1–3:10. ACM, NY, USA (2008). doi:10.1145/1408681.1408684
27. von Löwis, M., Denker, M., Nierstrasz, O.: Context-oriented programming: Beyond layers. In: Proceedings of the 2007 International Conference on Dynamic Languages, ICDL '07, pp. 143–156. ACM, NY, USA (2007). http://dx.doi.org/10.1145/1352678.1352688
28. Yin, Z., Ma, X., Zheng, J., Zhou, Y., Bairavasundaram, L.N., Pasupathy, S.: An empirical study on configuration errors in commercial and open source systems. In: Proceedings of the Twenty-Third ACM Symposium on Operating Systems Principles, SOSP '11, pp. 159–172. ACM, New York, NY, USA (2011). doi:10.1145/2043556.2043572
29. Yuan, D., Park, S., Zhou, Y.: Characterizing logging practices in open-source software. In: Proceedings of the 34th International Conference on Software Engineering, ICSE '12, pp. 102–112. IEEE Press, Piscataway, NJ, USA (2012). http://dl.acm.org/citation.cfm?id=2337223.2337236

Stripes-Based Object Matching

Oliver Tiebe, Cong Yang, Muhammad Hassan Khan, Marcin Grzegorzek and Dominik Scarpin

Abstract We propose a novel and fast 3D object matching framework that is able to fully utilise the geometry of objects without any object reconstruction process. Traditionally, 3D object matching methods are mostly applied based on 3D models. In order to generate accurate and proper 3D models, object reconstruction methods are used for the collected data from laser or time-of-flight sensors. Although those methods are naturally appealing, heavy computations are required for segmentation as well as transformation estimation. Moreover, some useful features could be filtered out during the reconstruction process. On the contrary, the proposed method is applied without any reconstruction process. Building on stripes generated from laser scanning lines, we represent an object by a set of stripes. To capture the full geometry, we describe each stripe by the proposed robust point context descriptor. After representing all stripes, we perform a flexible and fast matching over all collected stripes. We show that the proposed method achieves promising results on some challenging real-life objects.

Keywords Object matching · Stripe matching · Laser scanner · Point context

1 Introduction

Determining the similarity between 3D objects is a fundamental task for many robotic and industrial applications [1, 2] such as 3D shape retrieval, face morphing, and object recognition [3]. A challenging aspect of this task is to find suitable object signatures that can be constructed and compared quickly, while still discriminating

O. Tiebe · C. Yang (✉) · M.H. Khan · M. Grzegorzek
Institute for Vision and Graphics, University of Siegen, Hoelderlinstr. 3,
57076 Siegen, Germany
e-mail: cong.yang@uni-siegen.de

D. Scarpin
Institute of Automatic Control Engineering, University of Siegen, Hoelderlinstr. 3,
57076 Siegen, Germany
e-mail: scarpin@zess.uni-siegen.de

R. Lee (ed.), *Computer and Information Science*,
Studies in Computational Intelligence 656, DOI 10.1007/978-3-319-40171-3_5

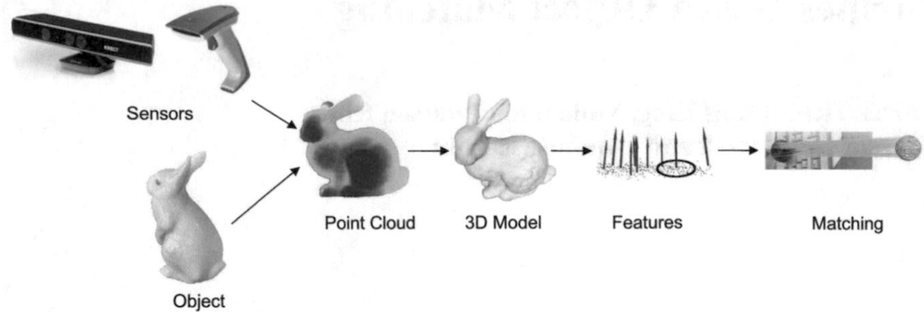

Fig. 1 Pipeline of the traditional object representation and matching

between similar and dissimilar objects. With traditional approaches [4–6], accurate and proper 3D models [7–9] are firstly reconstructed to feature the geometrical and textural properties of objects. Specifically, as shown in Fig. 1, the first step is the object scanning via time-of-flight cameras [10] or laser scanning systems [11]. After that, the scanned object is represented by some special formats (e.g. point cloud) [12] with a filtering process to remove outliers and noise. Finally, original surfaces from 3D scans are reconstructed into an object model using meshes or other formats. For object matching, 3D features are generated at a certain 3D point or position in space, which describe geometrical patterns based on the information available around the point. Finally, the similarity between two 3D objects is calculated based on object matching.

However, it shares some challenges with the pipeline of traditional approaches. For object reconstruction, some heavy computation costs may be required depending on the outliers in point clouds and tasks at hand. Specifically, Due to the background clutter and measurement errors, certain objects present a large number of shadow points. This complicates the estimation of local point cloud 3D features. Thus, some of these outliers should be filtered by performing a statistical analysis on each point's neighbourhood, and trimming those which do not meet a certain criteria [12]. This filtering process normally calls for a large number of calculations. Moreover, if the point cloud is composed of multiple scans that are not aligned perfectly, a smoothing and re-sampling process is also required. In addition, during the outlier removing process, some fine-grained features could be filtered out.

For object matching, in order to find reliable feature point correspondences, some high-order graph matching frameworks are employed to establish feature correspondences, combining both appearance similarity and geometric compatibility [13–15]. Although those methods have been successfully applied in 2D image features, limited prior art refers 3D surfaces. The main reason is that a 3D surface is not represented in the Euclidean 2D domain, and therefore distances between two points on the surface cannot be computed in a closed form [16]. Moreover, computing object similarity using correspondences normally requires high computational complexity

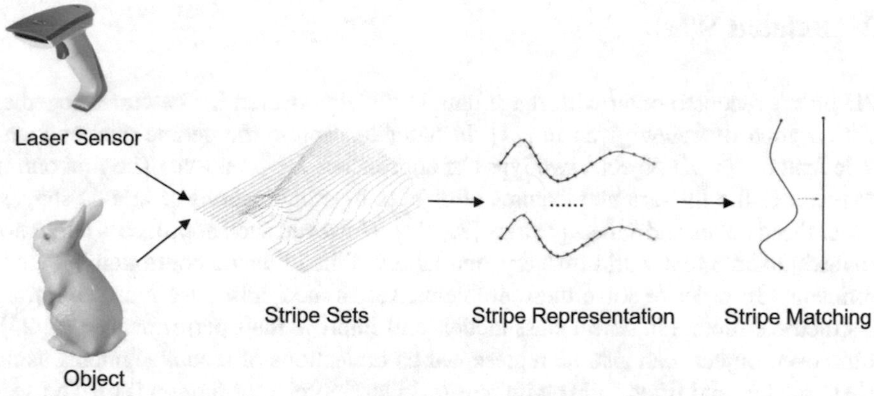

Fig. 2 Pipeline of the proposed representation and matching methods

(e.g. the most commonly used Hungarian algorithm [17] needs $O(n^3)$ time complexity, where n is the number of feature points) since each feature point in one object should be assigned to a point in another object.[1] Thus, it is hard to be applied in real time.

In order to solve the above problems, we present a method that is able to efficiently represent a 3D object using scanning stripes without any reconstruction process. With this, similarity between objects can be calculated directly using vector distance methods [19, 20] with low computational complexity. Specifically, as shown in Fig. 2, instead of object reconstruction using point clouds, we represent an object into a set of stripes which are collected from laser scanning lines. In order to capture the full geometry of an object, we process and describe each stripe by the proposed point context descriptor. After representing all stripes, we perform a flexible and fast matching over all collected stripes. Since the collected scripts are naturally ordered by the moving direction of a scanning laser, stripes can be easily matched based on their relative locations. With this property, our matching task is applied in real time without any heavy stripe corresponding process.

The most significant scientific contributions of this paper include: (1) We propose a novel and efficient stripe-based object representation method without any 3D object reconstruction process. (2) In order to fully capture the geometrical properties of each stripe, we introduce an intuitive and robust point context descriptor. (3) We introduce a fast matching method for calculating the similarity between two stripe sets. This approach is applied without any corresponding process and the matching complexity can be reduced dramatically. (4) The experiments show that some objects with similar shapes can be classified accurately using the proposed method.

[1] In some matching tasks, partial feature points could be jumped using dummy points like in [18].

2 Related Work

3D object reconstruction with range data is widely covered in literatures together with a good overview given in [21]. In order to capture the geometrical and surface features of 3D objects, two types of approaches are proposed. The first one is to preserve the fine-grained features of objects by combinations of atomic shapes, generalised cones and super-quadrics [22, 23]. However, such approaches could not robustly handle real world imagery, and largely failed outside controlled lab environments. In order to solve these problems, researchers introduce more and more geometric structure in object class models and improve their performance [24, 25]. Moreover, objects can also be represented as collections of planar segments using CAD models and lifted to 3D with non-rigid structure-from-motion [26]. The second one is to preserve the coarse-grained features of objects by combining multiple simple shapes to obtain object models [27]. This idea is further improved to the level of plane- and box-type models [28, 29]. Though most works [28–30] indicate that both fine- and coarse-grained models can help one to better guess the 3D layout of an object while at the same time improving 2D recognition, those methods normally require high computing time for processing and analysing 3D surfaces since most surfaces rarely have simple parametrisations. In addition, since 3D surfaces can have arbitrary topologies, many useful methods for analysing other media have no obvious equivalent for surface models. On the contrary, as we directly employ the scanning stripes for object representation, the proposed method is applied without any 3D reconstruction process.

For object matching, the biggest challenge is the large non-rigid deformations of object surfaces. In applications such as facial expression recognition, there are localised, high-degree of freedom deformations. To tackle this problem, two types of approaches are normally employed [16, 31]. The first one obtains dense feature point correspondences by embedding the surfaces to a canonical domain which preserves the geodesics or angles [32, 33]. Such embedding requires an initial set of feature correspondences or boundary conditions. However, it is difficult to find reliable feature point correspondences and consistent boundary conditions. Furthermore, since most surface deformations are not perfectly isometric, solely considering intrinsic embedding information may introduce approximation errors to the matching results. Therefore, Zeng et al. propose an approach to achieve robust dense surface matching via high-order graph matching in the embedding manifold [13, 16]. Specifically, they use multiple measurements to capture the appearance and geometric similarity between deformed surfaces and high-order graph interaction to model the implicit embedding energy. As these approaches require high computational complexity for optimising high-order graphs, they are hard to be applied in real time. The second type is to represent 3D models using skeletons and then skeleton matching approaches are employed for matching objects [18, 34]. However, as skeletonisation methods [31, 35] are normally sensitive to noise, the generated skeletons require an extra skeleton pruning process [36]. Moreover, similar to the feature point matching approaches, skeleton matching is built on skeleton graphs which require expensive computational time for search correspondences. Different from the aforementioned approaches, the proposed method calculates the similarity between objects without any corresponding process since the generated stripes are naturally ordered. Therefore, our method can be applied in real time.

3 Stripe Generation

In this section, the stripe generation method is introduced. The stripe generation is done with a robot guided 2D laser scanner (see Fig. 3a), which was developed for the modiCAS [37] project at the University of Siegen. The original purpose of this system is to assist medical personnel in a surgical environment to acquire the patients face anatomy, to perform intra-operative patient registration, surgical navigation and placement of medial tools.

As shown in Fig. 3b, the laser scanner is build up from a commercial 3D stereo-vision system, developed by the company Point Grey, which is normally used for marker based tracking. To extend this system to a high resolution laser scanner, a line laser module is rigidly attached to one of the two integrated cameras of the stereo-vision system. Afterwards a camera calibration is done to calculate the intrinsic camera parameters, which are needed to correct distortions caused by the camera lens. At least the laser scanner is calibrated with a special calibration device for correct distance measurement. The achieved measurement accuracy of the laser scanner is less than 0.3 mm.

For a measurement with the laser scanner the line laser module is used to project a laser line to the object surface and the reflection of the laser then is detected by the camera. Due to the characteristic of the objects surface, the laser line is deformed in the acquired camera image. From this deformation, and the knowledge of the camera

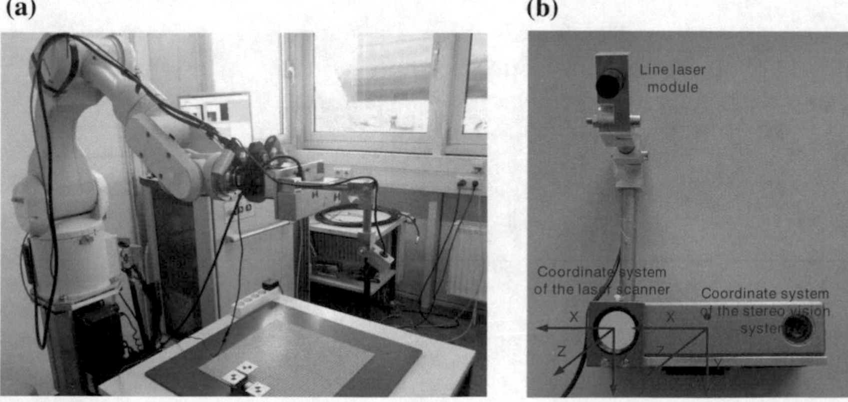

Fig. 3 The laser scanning system and its components. **a** The modiCAS system, **b** Components

parameters and the triangulation angle between the camera and the line laser module, the distance between the camera and the object surface can be calculated.

To ensure a solid detection of the reflected laser line in the camera image, it is necessary to avoid any influences from background light. To suppress any background light, an optical bandpass filter, which is designed to only let light around a wavelength of 650 nm pass through, is mounted in front of the camera lens used. With the help of such a filter the laser line is the brightest object in the camera image and can be detected easily. For this, in every column of the image array the start and the end of the laser line are detected by an adjustable threshold. Afterwards, the correct position of the laser line in each column is calculated using a weighted average of the intensity values of the laser between the threshold borders.

For the acquisition of the laser lines, the laser scanner is mounted to the robot arm with the help of a rigid fast coupling and a hand-eye calibration is done to calculate the transformation matrix between the coordinate systems of the optical system and the robot flange. Afterwards, the objects are placed one after another on a table, with a distance of approximately 60 cm between the table and the laser scanner, and the laser scanner is moved with the robot arm along the object's surface with a homogenous speed, so the distance between the laser lines is constant over the whole object's surface. The density of the laser lines on the object surface can be affected by the moving speed of the robot arm.

After completion of the data acquisition the detected laser lines are saved in an array, which is transferred from the camera control PC to the user PC. For further processing, the found line positions are recalculated to black/white images. The calculated distance values are not needed for this project. Figure 4 shows an example of the original object (Fig. 4a) and its collected stripes (Fig. 4c). For comparison, a 3D model of the original object is illustrated in Fig. 4b. We can observe that the proposed stripe descriptor is more simple than the 3D model. Moreover, considering the generation speed, the proposed descriptor is much faster than 3D models. Here we only illustrate the stripes from one side of the surface. To capture the full geometry, we can conduct multiple scans while changing the object's pose. However, experi-

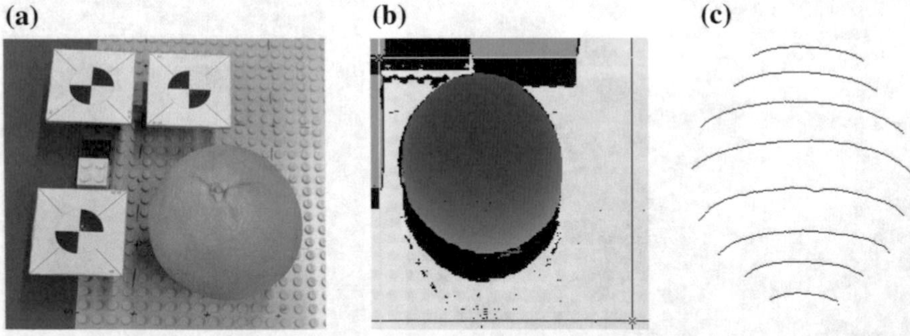

Fig. 4 An apple and its 3D model and generated stripes (partly). **a** Original object, **b** 3D model, **c** Collected strips

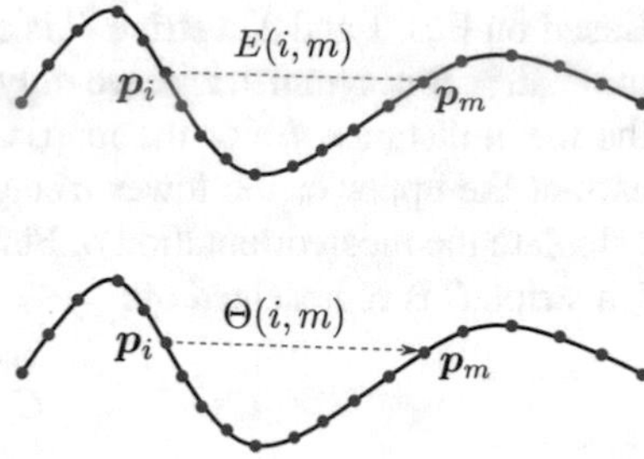

Fig. 5 The proposed descriptor for a stripe C

ments in Sect. 6 show that even with the single-side stripes, matching performances are still promising. Thus, the complete scanning strategy is applied based on different applications.

4 Stripe Representation

For a given stripe, we describe its geometrical and topological properties by the point context descriptor. Specifically, given a stripe $\boldsymbol{C}$ with H points, for every point $\boldsymbol{p}_i \in \boldsymbol{C}, i = 1, 2, \ldots, H$, we consider both the distance and direction of the vector form $\boldsymbol{p}_i$ to other points in $\boldsymbol{C}$. Then, the mean distance and direction are calculated for stripe representation. Moreover, in order to distinguish the straight line-similar stripes, the normalised[2] stripe length l is also employed for stripe description. Thus, a stripe is represented by a three-dimensional feature vector. The proposed descriptor has many characteristics: (1) It is simple and intuitive. (2) It integrates both geometrical and topological features of a stripe. (3) It is flexible for stripe matching since it can be adopted to different matching algorithms.

More specifically, as shown in Fig. 5, given a stripe $\boldsymbol{C}$ with point sequence $\boldsymbol{C} = \boldsymbol{p}_1, \boldsymbol{p}_2, \ldots, \boldsymbol{p}_H$, we compute two matrices, one presenting all distances and the second one representing all pairwise orientations of vectors from each $\boldsymbol{p}_i$ to each $\boldsymbol{p}_m \in \boldsymbol{C}, m = 1, 2, \ldots, H$. The distance $E(i, m)$ from $\boldsymbol{p}_i$ to $\boldsymbol{p}_m$ is defined as the Euclidean distance in the log space:

$$E(i, m) = \log(1 + \|\overrightarrow{\boldsymbol{p}_i} - \overrightarrow{\boldsymbol{p}_m}\|_2) \quad . \tag{1}$$

We add one to the Euclidean distance to make the $E(i, m)$ positive. The orientation $\Theta(i, m)$ from $\boldsymbol{p}_i$ to $\boldsymbol{p}_m$ is defined as the orientation of vector $\overrightarrow{\boldsymbol{p}_i} - \overrightarrow{\boldsymbol{p}_m}$:

$$\Theta(i, m) = \angle(\overrightarrow{\boldsymbol{p}_i} - \overrightarrow{\boldsymbol{p}_m}) \in [-\pi, \pi] \quad . \tag{2}$$

[2]Here we normalise a stripe length H by the mean length of all stripes in an object.

Based on Eqs. 1 and 2, a stripe $\boldsymbol{C}$ is encoded in two $H \times H$ matrices E and Θ. Since the matrix E is symmetrical, we only extract its lower triangular part for calculating the mean distance d. For the matrix Θ, as its absolute values are symmetrical, we extract the upper or the lower triangular part which has more positive values and calculate the mean orientation o. Finally, together with the normalised stripe length l, a stripe $\boldsymbol{C}$ is represented by:

$$\boldsymbol{C} = [l, d, o] \quad . \tag{3}$$

5 Stripe Matching

Let $\boldsymbol{A}_1$ and $\boldsymbol{A}_2$ denote sets of stripes from two objects O_1 and O_2, respectively. $\boldsymbol{C}_i$ and $\boldsymbol{C}'_j$ denote a single stripe in $\boldsymbol{A}_1$ and $\boldsymbol{A}_2$, $i = 1, 2, \ldots, N, j = 1, 2, \ldots, M$. For notational simplicity we assume that $N \leqslant M$. Our aim is to calculate the similarity between O_1 and O_2 using their stripe sets. As each stripe is represented by a three-dimensional feature vector, we calculate the similarity between O_1 and O_2 using the properties of each feature distribution.

Specifically, for two objects O_1 and O_2, we first remove some redundant stripes from $\boldsymbol{A}_2$ to ensure they have the same number of stripes N. In order to do so, we remove the rounding number $(M - N)/2$ stripes from two ends of $\boldsymbol{A}_2$. For example, we remove the stripes $\{\boldsymbol{C}'_1, \ldots, \boldsymbol{C}'_{(M-N)/2}\}$ and the stripes $\{\boldsymbol{C}'_{N-((M-N)/2)+1}, \ldots, \boldsymbol{C}'_N\}$. The rationale behind this is (1) Stripes which are close to the boundary have less influence on the global structure of an object. (2) Most objects have symmetrical structures. (3) This strategy can avoid the removing of some crucial stripes which have major contribution for object distinction. With the above step, $\boldsymbol{A}_1$ and $\boldsymbol{A}_2$ have the same number of stripes N. With the original order, we renumber the index of stripes in $\boldsymbol{A}_2$ as:

$$\boldsymbol{A}_2 = \{\boldsymbol{C}'_1, \boldsymbol{C}'_2, \ldots, \boldsymbol{C}'_N\} \quad . \tag{4}$$

As each stripe can be represented by a three-dimensional feature vector $\boldsymbol{C}_i = [l_i, d_i, o_i]$, $\boldsymbol{C}'_i = [l'_i, d'_i, o'_i]$, we capture the distribution of each feature by its feature values in all stripes. Then the distance between O_1 and O_2 can be calculated by Bhattacharyya distance [20]. Specifically, let $\boldsymbol{l}$, $\boldsymbol{d}$ and $\boldsymbol{o}$ denote the distributions of all feature values in $\boldsymbol{A}_1$. $\boldsymbol{l}'$, $\boldsymbol{d}'$ and $\boldsymbol{o}'$ denote the distributions of all feature values in $\boldsymbol{A}_2$. For example, $\boldsymbol{l} = [l_i, l_2, \ldots, l_N]$ and $\boldsymbol{l}' = [l'_1, l'_2, \ldots, l'_N]$. Assume $\boldsymbol{K} = \{\boldsymbol{l}, \boldsymbol{d}, \boldsymbol{o}\}$ and $\boldsymbol{K}' = \{\boldsymbol{l}', \boldsymbol{d}', \boldsymbol{o}'\}$, the distance between O_1 and O_2 is calculated by

$$s(O_1, O_2) = \tfrac{1}{3} \sum_{t=1}^{3} \lambda_t D_B(\boldsymbol{K}(t), \boldsymbol{K}'(t)) \tag{5}$$

where λ is the weight for fusing three features and $D_B(\boldsymbol{K}(t), \boldsymbol{K}'(t))$ denotes the Bhattacharyya distance between two feature distributions, e.g. $D_B(\boldsymbol{K}(1),$

$\boldsymbol{K}'(1)) = D_B(\boldsymbol{l}, \boldsymbol{l}')$. In practice, λ can be searched using the heuristic method of Gradient Hill Climbing integrated with Simulated Annealing [38]. Specifically, the Gradient Hill Climbing [39] method starts with randomly selected parameters. Then it changes single parameters iteratively to find a better set of parameters. A fitness function then evaluates whether the new set of parameters performs better or worse. The Simulated Annealing strategy [40] impacts the degree of the changes. In later iterations, the changes to the parameters are becoming smaller. With our preliminary experiments, we set $\lambda_1 = 0.6$, $\lambda_2 = 0.3$ and $\lambda_3 = 0.1$ for three features. Furthermore, the Bhattacharyya distance $D_B(\boldsymbol{l}, \boldsymbol{l}')$ is calculated by:

$$D_B(\boldsymbol{l}, \boldsymbol{l}') = \frac{1}{4}\ln(\frac{1}{4}(\frac{\sigma_l^2}{\sigma_{l'}^2} + \frac{\sigma_{l'}^2}{\sigma_l^2} + 2)) + \frac{1}{4}(\frac{(\mu_l - \mu_{l'})^2}{\sigma_l^2 + \sigma_{l'}^2}) \tag{6}$$

where σ and μ are the variance and mean of a feature distribution, respectively. Similar to $D_B(\boldsymbol{K}(1), \boldsymbol{K}'(1))$, $D_B(\boldsymbol{K}(2), \boldsymbol{K}'(2)) = D_B(\boldsymbol{d}, \boldsymbol{d}')$ and $D_B(\boldsymbol{K}(3), \boldsymbol{K}'(3)) = D_B(\boldsymbol{o}, \boldsymbol{o}')$ can also be calculated with Eq. 6.

6 Experiments

In this section we first introduce the dataset we used for the experiments. After that, we evaluate and compare the performance of the proposed method with some traditional methods to illustrate our advantages. Lastly, we analyse the computational complexity of the proposed method. The experiments in this paper are performed on a laptop with Inter Core i7 2.2 GHz CPU, 8.00 GB memory and 64-bit Windows 8.1 OS. All methods in our experiments are implemented in Matlab R2015a.

6.1 Dataset

To validate the idea of our proposed method on real-life objects, we organised a dataset namely Daily100 from daily life. The Daily100 database includes 100 objects with 10 classes, such as apple, banana, book, chips, chopstick, egg, orange, pear, pen and bottle cap (the first row in Fig. 6). For each object, we also generated and collected its correlated stripes and shape for the experiment. From Fig. 6 we can see that some objects (e.g. apple and orange) are really difficult to be distinguished using only their shapes.

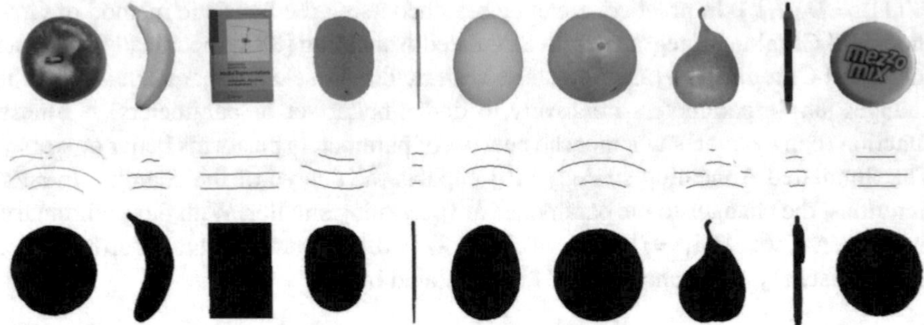

Fig. 6 Sample objects of the proposed Daily100 dataset. The *first row* illustrates the original objects. The *second* and *third row* show the sample stripes and shape of each object, respectively

6.2 Experiment Performance

Based on the Daily100 dataset, we perform two experiments within the object retrieval frame work. Specifically, in the first experiment, we compare the global object retrieval performances between stripes and shapes. In the second experiment, more detailed comparisons in each object class are illustrated and discussed.

Table 1 depicts the matching performance of the proposed method and other shape-based approaches. We use each object as a query and retrieve the 10 most similar objects among the whole dataset. The final value in each position is counter values that are obtained by checking retrieval results using all the 100 objects as queries. For example, the fourth position in the row of our method shows that from 100 retrieval results in this position, 89 objects have the same class as the query objects. We can clearly observe that the proposed method achieves the best results among all the other methods. The main reason is that since most objects have similar shapes (e.g. apple, orange and bottle cap), it is hard to distinguish them using only their shape features. Different from the shape features, the proposed method captures and preserves deformations on object surfaces using stripe sets. Thus, the proposed method can distinguish objects even if their shapes are similar.

Table 1 Object retrieval comparison between the proposed method and shape-based approaches: Shape Context (SC), Inner Distance (ID) and Path Similarity (PS)

	1	2	3	4	5	6	7	8	9	10
SC [41]	100	79	84	71	75	74	64	67	58	48
ID [42]	100	80	77	70	68	68	64	66	54	41
PS [18]	100	81	79	78	76	71	65	60	55	53
HF [43]	100	80	71	67	64	54	52	43	49	51
Our	**100**	**95**	**91**	**89**	**86**	**77**	**73**	**72**	**66**	**63**

Table 2 Comparison of mean matching accuracy (%) in each class

	Apple	Banana	Book	Chips	Chopstick	Egg	Orange	Pear	Pen	Cap
SC [41]	46	100	79	73	87	77	51	79	100	28
HF [43]	53	82	48	61	51	43	54	97	100	30
Our	**100**	**78**	**48**	**100**	**100**	**96**	**84**	**100**	**42**	**64**

In order to perform a more detailed analysis of object matching, we report the mean matching accuracy in each object class in Table 2. In this table, the mean accuracy is calculated by the matching accuracy on each query object. Specifically, we use an object as the query and retrieve the 10 most similar objects among the whole dataset. Within these 10 objects, we count how many objects have the same class as the query object. The matching accuracy of the query object is then calculated by the ratio between the matched objects and 10. We use all the 100 objects as queries and calculate the mean accuracy for each class. In Table 2, we can observe that for most classes, our method achieves the best accuracy. For example, since apple, egg, orange and bottle cap have a very similar shape, the performances of shape-based methods in these classes are not promising. As the collected stripe sets have different geometrical properties in these classes, our method achieves the best performance. However, considering the classes of banana and pen, our method performs worse than the shape-based approaches. The main reason is that since both banana and pen have very similar surface deformation, the proposed stripe-based approach cannot robustly distinguish them using only stripes. In order to improve the accuracy, a proper combination of shape descriptors and the proposed stripe descriptor can improve the matching accuracy over the individual descriptor.

6.3 Computational Complexity and Runtime

We now analyse the computational complexity of the proposed hierarchical skeleton generation and matching approaches. (1) For stripe set generation, the time complexity is in the order of $O(Nl)$, where N is the number of stripes on each object and l is the mean stripe length. This is because our stripe can be directly generated from laser scanning lines. Thus, for each stripe, we only need $O(l)$ for thinning and noise removing. (2) For stripe representation, since we generated the point context descriptor using sample points along the stripe path, the time complexity is $O(H^2)$ where H is the number of sample points. Considering there are multiple stripes for each object, the global complexity for object representation is $O(NH^2)$. (3) For object matching, as we directly employ the Bhattacharyya distance on three-dimensional vectors, the complexity is $O(3)$. However, for each dimension, we need to calculate the mean value on the distance and orientation matrix, the global complexity for object matching is $O(3H)$. Thus, the total complexity of our method is

$O(Nl) + O(NH^2) + O(3N)$. By dropping the constant number, our time complexity is bounded by $O(Nl) + O(NH^2)$.

Here we report the computation time based on the Daily100 dataset with the experimental environment introduced above. On average, the shape resolution in this dataset is 600×712. For each object, the mean stripe number is 140. Together with object representation and matching, the proposed method takes 0.0375 h while the Shape Context [41] method takes 1.6223 h. Thus, our method can dramatically reduce the runtime while achieving promising matching accuracy. However, please notice that our code is not optimised, and its faster implementation is possible by optimising loops, settings and programming language, etc. Thus, there are still plenty of opportunities to reduce the running time.

7 Conclusion and Future Work

A novel 3D object matching method based on the similarity of stripes is presented. The most significant contribution of this paper is the novel approach to 3D object matching. We represent an object as a set of stripes which are directly collected from laser scanning lines. The distance between objects is computed using the Bhattacharyya distance based on stripe features. The proposed approach does not require any complicated strategies for 3D object reconstruction as well as the feature point corresponding. Thus, our method can dramatically reduce computational complexity for 3D object matching. In addition to low time costs, our method achieves a promising performance on some challenging objects compared to the traditional 2D shape-based approaches. In the future, we will try to optimise our stripe collection strategy for adapting hand-hold laser scanning devices. Moreover, we will compare our approach to other 3D object matching methods.

Acknowledgments Research activities leading to this work have been supported by the China Scholarship Council (CSC) and the German Research Foundation (DFG) within the Research Training Group 1564 (GRK 1564).

References

1. Drost, B., Ilic, S.: Graph-based deformable 3d object matching. Pattern Recognit. **9358**, 222–233 (2015)
2. Osada, R., Funkhouser, T., Chazelle, B., Dobkin, D.: Matching 3d models with shape distributions. In: International Conference on Shape Modeling and Applications, pp. 154–166 (2001)
3. Campbell, R.J., Flynn, P.J.: A survey of free-form object representation and recognition techniques. Comput. Vis. Image Underst. **81**(2), 166–210 (2001)
4. Hong, C., Yu, J., You, J., Chen, X., Tao, D.: Multi-view ensemble manifold regularization for 3d object recognition. Inf. Sci. **320**, 395–405 (2015)
5. Leng, B., Zeng, J., Yao, M., Xiong, Z.: 3d object retrieval with multitopic model combining relevance feedback and lda model. IEEE Trans. Image Process. **24**(1), 94–105 (2015)

6. Yu, Y., Li, J., Guan, H., Jia, F., Wang, C.: Three-dimensional object matching in mobile laser scanning point clouds. IEEE Geosci. Remote Sens. Lett. **12**(3), 492–496 (2015)
7. Bernardini, F., Bajaj, C.L., Chen, J., Schikore, D.R.: Automatic reconstruction of 3d cad models from digital scans. Int. J. Comput. Geom. Appl. **9**, 327–369 (1999)
8. Maier, R., Sturm, J., Cremers, D.: Submap-based bundle adjustment for 3d reconstruction from rgb-d data. Pattern Recognit. **8753**, 54–65 (2014)
9. Kehl, W., Navab, N., Ilic, S.: Coloured signed distance fields for full 3d object reconstruction. In: Proceedings of the British Machine Vision Conference. BMVA Press (2014)
10. Lefloch, D., Nair, R., Lenzen, F., Schfer, H., Streeter, L., Cree, M., Koch, R., Kolb, A.: Technical foundation and calibration methods for time-of-flight cameras. In: Time-of-Flight and Depth Imaging. Sensors, Algorithms, and Applications. Springer Berlin Heidelberg, vol. 8200, pp. 3–24 (2013)
11. Kedzierski, M., Fryskowska, A.: Methods of laser scanning point clouds integration in precise 3d building modelling. Measurement **74**, 221–232 (2015)
12. Rusu, R., Cousins, S.: 3d is here: Point cloud library (pcl). In: IEEE International Conference on Robotics and Automation, pp. 1–4 (2011)
13. Duchenne, O., Bach, F., Kweon, I.-S., Ponce, J.: A tensor-based algorithm for high-order graph matching. IEEE Trans. Pattern Anal. Mach. Intell. **33**(12), 2383–2395 (2011)
14. Leordeanu, M., Hebert, M.: A spectral technique for correspondence problems using pairwise constraints. In: IEEE International Conference on Computer Vision, vol. 2, pp. 1482–1489 (2005)
15. Yang, C., Feinen, C., Tiebe, O., Shirahama, K., Grzegorzek, M.: Shape-based object matching using point context. In: International Conference on Multimedia Retrieval, pp. 519–522 (2015)
16. Zeng, Y., Wang, C., Wang, Y., Gu, X., Samaras, D., Paragios, N.: Dense non-rigid surface registration using high-order graph matching. In: IEEE Conference on Computer Vision and Pattern Recognition, pp. 382–389 (2010)
17. Jonker, R., Volgenant, T.: Improving the hungarian assignment algorithm. Oper. Res. Lett. **5**(4), 171–175 (1986)
18. Bai, X., Latecki, L.: Path similarity skeleton graph matching. IEEE Trans. Pattern Anal. Mach. Intell. **30**(7), 1282–1292 (2008)
19. Breu, H., Gil, J., Kirkpatrick, D., Werman, M.: Linear time euclidean distance transform algorithms. IEEE Trans. Pattern Anal. Mach. Intell. **17**(5), 529–533 (1995)
20. Dubuisson, S.: The computation of the bhattacharyya distance between histograms without histograms. In: International Conference on Image Processing Theory Tools and Applications, pp. 373–378 (2010)
21. Bernardini, F., Rushmeier, H.: The 3d model acquisition pipeline. Comput. Graph. Forum **21**(2), 149–172 (2002)
22. Brooks, R.A.: Symbolic reasoning among 3-d models and 2-d images. Artif. Intell. **17**(1), 285–348 (1981)
23. Pentland, A.P.: Perceptual organization and the representation of natural form. Artif. Intell. **28**(3), 293–331 (1986)
24. Leibe, B., Leonardis, A., Schiele, B.: An implicit shape model for combined object categorization and segmentation. In: Towards Category-Level Object Recognition, pp. 496–510 (2006)
25. Stark, M., Goesele, M., Schiele, B.: Back to the future: Learning shape models from 3d cad data. In: British Machine Vision Conference, pp. 1–11 (2010)
26. Savarese, S.: Estimating the aspect layout of object categories. In: IEEE Conference on Computer Vision and Pattern Recognition, pp. 3410–3417 (2012)
27. Roberts, L.G.: Machine perception of three-dimensional soups. Ph.D. dissertation, Massachusetts Institute of Technology (1963)
28. Silberman, N., Hoiem, D., Kohli, P., Fergus, R.: Indoor segmentation and support inference from rgbd images. In: European Conference on Computer Vision, pp. 746–760 (2012)
29. Liu, X., Zhao, Y., Zhu, S.-C.: Single-view 3d scene parsing by attributed grammar. In: IEEE Conference on Computer Vision and Pattern Recognition, pp. 684–691 (2014)

30. Zia, M.Z., Stark, M., Schindler, K.: Towards scene understanding with detailed 3d object representations. Int. J. Comput. Vis. **112**(2), 188–203 (2014)
31. Tagliasacchi, A., Zhang, H., Cohen-Or, D.: Curve skeleton extraction from incomplete point cloud. ACM Trans. Graph. **28**(3), 1–9 (2009)
32. Wang, S., Wang, Y., Jin, M., Gu, X.D., Samaras, D.: Conformal geometry and its applications on 3d shape matching, recognition, and stitching. IEEE Trans. Pattern Anal. Mach. Intell. **29**(7), 1209–1220 (2007)
33. Zeng, W., Zeng, Y., Wang, Y., Yin, X., Gu, X., Samaras, D.: 3d non-rigid surface matching and registration based on holomorphic differentials. In: European Conference on Computer Vision, pp. 1–14 (2008)
34. Yang, C., Tiebe, O., Shirahama, K., Grzegorzek, M.: Object matching with hierarchical skeletons. Pattern Recognit. (2016)
35. Huang, H., Wu, S., Cohen-Or, D., Gong, M., Zhang, H., Li, G., Chen, B.: L1-medial skeleton of point cloud. ACM Trans. Graph. **32**(4), 1–8 (2013)
36. Bai, X., Latecki, L.J., Liu, W.-Y.: Skeleton pruning by contour partitioning with discrete curve evolution. IEEE Trans. Pattern Anal. Mach. Intell. **29**(3), 449–462 (2007)
37. Cruces, R.A.C., Schneider, H.C., Wahrburg, J.: Cooperative robotic system to support surgical interventions. INTECH Open Access Publisher, pp. 481–490 (2008)
38. Yang, C., Tiebe, O., Pietsch, P., Feinen, C., Kelter, U., Grzegorzek, M.: Shape-based object retrieval and classification with supervised optimisation. In: International Conference on Pattern Recognition Applications and Methods, pp. 204–211 (2015)
39. Russell, S., Norvig, P.: Artificial Intelligence: A Modern Approach, 3rd edn. Prentice Hall Press (2009)
40. Kirkpatrick, S., Gelatt, C.D., Vecchi, M.P.: Optimization by simulated annealing. Science, pp. 671–680 (1983)
41. Belongie, S., Malik, J., Puzicha, J.: Shape matching and object recognition using shape contexts. IEEE Trans. Pattern Anal. Mach. Intell. **24**(4), 509–522 (2002)
42. Ling, H., Jacobs, D.: Shape classification using the inner-distance. IEEE Trans. Pattern Anal. Mach. Intell. **29**(2), 286–299 (2007)
43. Wang, J., Bai, X., You, X., Liu, W., Latecki, L.J.: Shape matching and classification using height functions. Pattern Recognit. Lett. **33**(2), 134–143 (2012)

User Engagement Analytics Based on Web Contents

Phoom Chokrasamesiri and Twittie Senivongse

Abstract User engagement is a relation of emotion, cognitive, and behavior between users and resources at a specific time or range of time. Measuring and analyzing web user engagement has been used by web developers as a means to gather feedback information from web users in order to understand their behavior and find ways to improve the websites. Many websites have been successful in using analytics tools since the information acquired by the tools helps, for example, to increase sales and the rate of returning to the websites. Most web analytics tools in the market focus on measuring engagement with the whole webpages, whereas the insight information about user behavior with respect to particular contents or areas within webpages is missing. However, such knowledge of web user engagement based on contents of the webpages would provide a deeper perspective on user behavior, compared to that based on the whole webpages. To fill this gap, we propose a set of web-content-based user engagement metrics that are adapted from existing web-page-based engagement metrics. In addition, the proposed metrics are accompanied by an analytics tool which the web developers can install on their websites to acquire deeper user engagement information.

Keywords User engagement · Web analytics · Web contents · Metric

1 Introduction

User Engagement is a relation of emotion, cognitive, and behavior between users and resources at a specific time or range of time [1]. In the past years, user engagement analytics on websites has been widely used by web developers to gather feedback from web users. In a survey of datanyze.com, 97.6 % of Alexa top one

P. Chokrasamesiri · T. Senivongse (✉)
Department of Computer Engineering, Faculty of Engineering,
Chulalongkorn University, Bangkok 10330, Thailand
e-mail: twittie.s@chula.ac.th

P. Chokrasamesiri
e-mail: phoom.c@student.chula.ac.th

R. Lee (ed.), *Computer and Information Science*,
Studies in Computational Intelligence 656, DOI 10.1007/978-3-319-40171-3_6

million websites use web analytics tools to analyze their user behavior data to obtain the insight of web usage [2]. The example of the analytics tools with the highest market share are Google Analytics, Google Universal Analytics, Yandex Metrica, comScore, and Quantcast, where Google Analytics occupies 45.5 % among these tools.

The analytics tools can provide web developers with the insight and feedback of web usage including what the users do on the websites, when they visit, how they interact with the websites, and many more. Once equipped with such information, the web developers can find ways to improve their web sites. For example, using an analytics tool, an e-commerce company [3] could identify the lost revenue due to high shopping cart abandonment rate and improved its website features to finally obtain an increase in checkout to payment page. Another case of a company [4], using an analytics tool to analyze which versions of its landing page should be used, could obtain an increase in homepage engagement and a boost to e-commerce conversion rate.

To analyze user engagement, a number of metrics have been developed to measure web usage and calculate user engagement. Click-Through Rate, Time Spent on a Site, Page Views, Return Rates, and Number of Unique Users are examples of popular metrics [5]. However, available analytics tools and metrics focus on analytics of user engagement with the whole webpages, rather than the web contents. Figure 1 depicts web contents, namely c1, c2, and c3, on a webpage, where Fig. 1a shows page-based engagement data that can be collected from users, i.e. mean visit to a page, mean hover over a page, and mean click to a page. However, the usage information regarding a particular content on a page is not known. If we can enhance page-based engagement metrics with content-based ones, i.e. mean visit to a con-

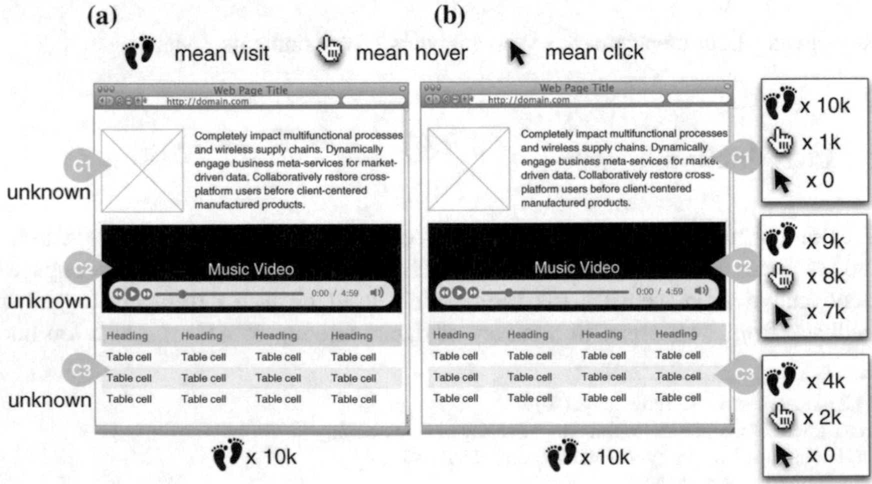

Fig. 1 Web usage analytics based on **a** webpage, **b** web contents on a page

tent, mean hover over a content, and mean click to a content as in Fig. 1b, web developers can obtain deeper understanding of how users interact with the contents and can improve their websites. This paper proposes a set of web-content-based user engagement metrics that are adapted from existing web-page-based engagement metrics. In addition, the proposed metrics are accompanied by an analytics tool which the web developers can install on their websites to acquire deeper user engagement information.

The rest of the paper is organized as follows. Section 2 discusses related work. Section 3 explains the web contents model and content-based user engagement measurement method. In Sect. 4, user engagement metrics based on web contents are presented together with a case study. Section 5 describes the content-based user engagement analytics tool, and the paper concludes in Sect. 6.

2 Related Work

We focus the discussion about related work on user engagement metrics and a popular web analytics tool.

2.1 Web User Engagement Metrics: A Web Analytics Approach

Lehmann et al. [6] study models of user engagement with web applications which can vary as the way users engage with different applications can be very different. A set of metrics, grouped into popularity, activity, or loyalty categories, is used in their study as listed in Table 1. These metrics are widely used to measure how users engage with webpages or websites. This paper adapts from these metrics to devise a set of web-content-based engagement metrics.

2.2 Web User Engagement Metrics: A Physiological Approach

A physiological approach to user engagement measures the users involuntary body responses, e.g. eye tracking, mouse tracking, and facial expression analysis [7]. The measurement may involve specific hardware device, e.g. eye tracking, or only software, e.g. mouse movement detection. In this paper, we consider only the metrics that can be measured by software, i.e. mouse gestures, in particular. Huang et al. [8] examine the users' mouse cursor behavior on search engine results pages to help better design effective search systems. The mouse cursor behavior includes clicks, cursor movements, and hovers over different page regions. We adapt from the metrics in Table 2 which are used in their work.

Table 1 User engagement metrics based on web analytics [6]

Category	Metric	Description
Popularity (for a given time frame)	Users	Number of distinct users
	Visits	Number of visits (one user can visit more than once)
	Clicks (page views)	Number of page views (measured by number of clicks on link)
Activity	ClickDepth	Average number of page views per visit
	DwellTimeA	Average time per visit
Loyalty (for a given time frame)	ActiveDays	Number of days a user visited website
	ReturnRate	Number of times a user visited website
	DwellTimeL	Average time a user spends on website

Table 2 User engagement metrics based on physiological measure [8]

Metric	Description
Clickthrough rate	Average number of link clicked per visit
Hover rate	Average number of link hovered (mouse over link) per visit
Number of unclicked hover	Median of number of unclicks on link that is hovered
Maximum hover time	Time that the link is hovered
Cursor trail length	Distance that cursor moves (pixel)
Movement time	Time that cursor moves (pixel)
Cursor speed	Cursor trail length/Movement time

2.3 Web Analytics Tool

We discuss Google Analytics [9] as an example of the tools which can be installed on a website to track how users engage with the website. Table 3 shows metrics that are used.

Table 3 User engagement metrics used by Google Analytics [9]

Metric	Description
Session	Total number of sessions within the date range
Users	Users that have had at least 1 session within the selected date range
Pageviews	Total number of pages viewed
Unique pageviews	Number of sessions during which the specified page was viewed at least once
Pages/session	Average number of pages viewed during a session
Average page duration	Average length of a session
Average time on page	Average amount of time users spent viewing pages
Bounce rate	Percentage of single-page visits (i.e. visitors left your site from the entrance page without interacting with the page)
% New session	An estimate of the percentage of first time visits
Entrances	Number of times visitors entered your site through a specified page or set of pages
% Exit	(Number of exits)/(Number of pageviews) for the page or set of pages

Google Analytics visualizes the measurements as graphs and tables. Figure 2 depicts how Google Analytics, for example, shows pageviews of a single page in a graph and others in a table. This paper will use similar visualization techniques to report web-content-based user engagement information.

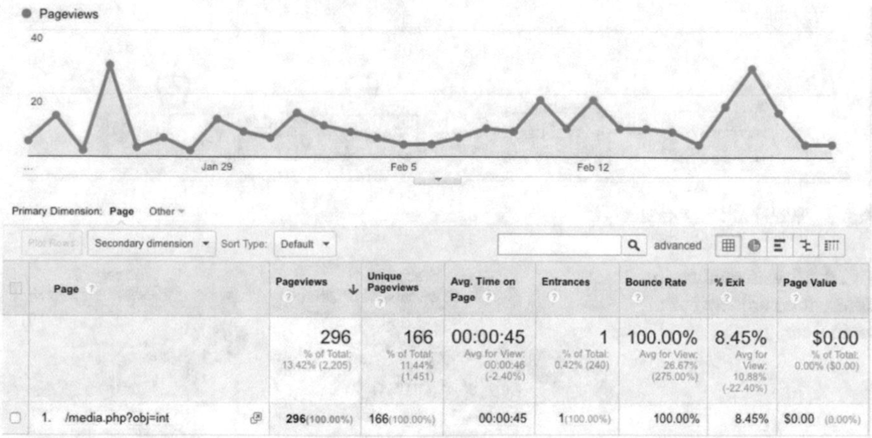

Fig. 2 Example of Google Analytics visualization

3 Web Contents Model and Content-Based Measurement

In this section, we give a definition of the web contents model, characteristics of web contents, and how user engagement is measured in this model.

3.1 Web Contents Model

The web contents model is depicted in Fig. 3. A website is a collection of webpages, where each webpage is composed of multiple containers. A container is a specific area on a particular webpage and can be nested inside another container. A web content is contained in a container, and can be seen on a web browser, e.g. text, image, sound, video etc. It can also be part of another content, e.g. text on an image.

3.2 Characteristics of Web Contents

Analyzing user engagement with web contents is different from that with webpages because of some unique characteristics of web contents as follows.

First, a web content can appear in different containers on different webpages at the same time. As shown in Fig. 4, a web content A can be in different containers on webpages 1 and 2 at the same time. We name a web content as cc or content in a container. That is, the same web content in different containers would be referred to

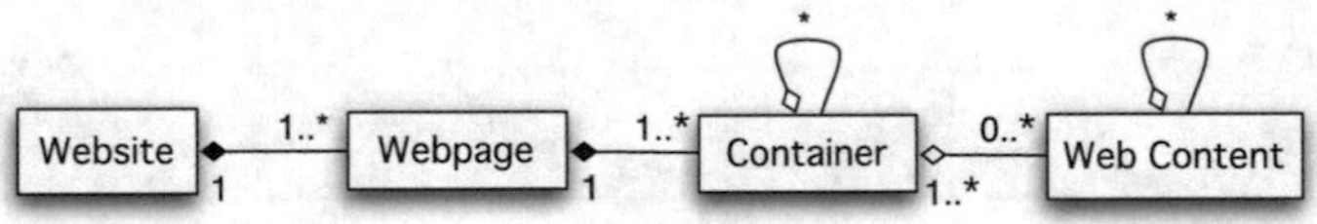

Fig. 3 Web contents model

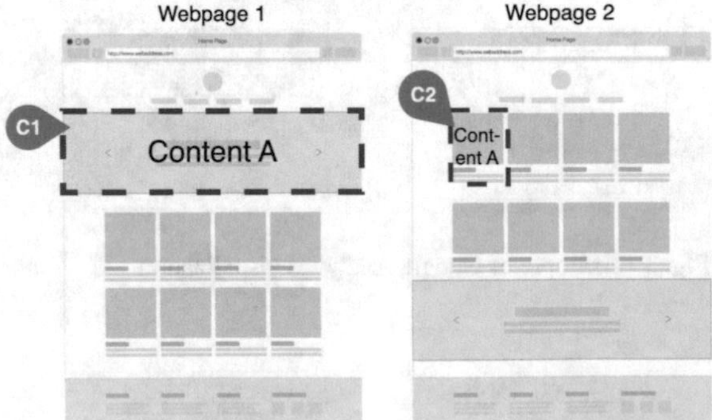

Fig. 4 A web content in different containers at the same time

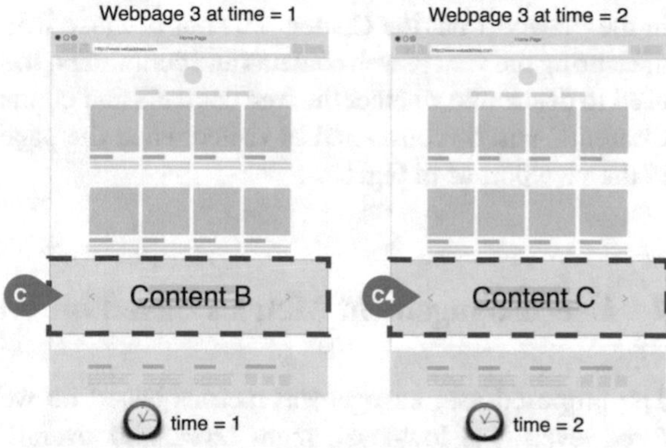

Fig. 5 Different web contents in the same container at different time

differently, e.g. the web content A in Fig. 4 is referred to as c1 and c2 with regard to its container on a webpage 1 and another one on a webpage 2 respectively.

Second, a container on a page can contain different web contents at different time. As shown in Fig. 5, a web content B contained in a container on a webpage 3 at time = 1 is referred to as c3. Later at time = 2, that container contains a different content C which will be referred to differently as c4.

Given these characteristics of web contents, we require a different set of metrics and tool for web-contents-based engagement analytics.

3.3 *Measuring Content and Container Visit*

Measuring visit to web contents and containers is different from measuring visit to webpages since the latter can be done when the page is loaded. However, for web contents and containers, they might be at the bottom of the page and cannot be seen by the user when the page is loaded. As shown in Fig. 6a, only the Content 1 is shown

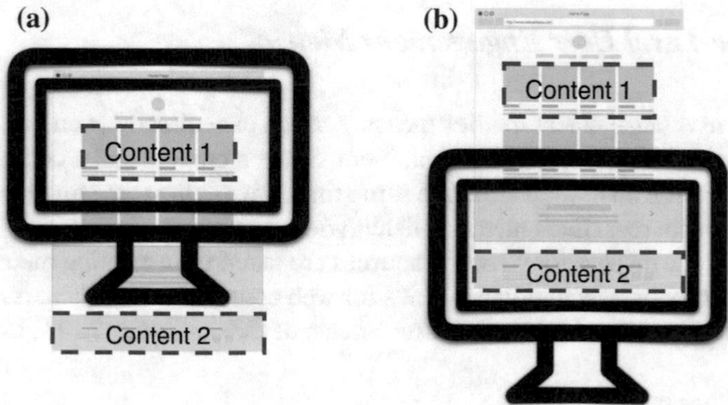

Fig. 6 Measuring content and container visit

in the viewport but the Content 2 is not because it is off the screen. In the case of measuring the visit to web contents and containers, the viewport of the screen will be used to determine whether the web contents and containers are visited. For example, Content 2 will be considered as visited when the page is scrolled down until shown in the viewport as in Fig. 6b.

4 User Engagement Metrics Based on Web Contents

The proposed user engagement metrics based on web contents are classified into three levels, i.e. low-level, high- level, and overall-level metrics. The low-level metrics are used to measure directly the behavior of the users on web contents. High-level metrics then are derived from the low-level ones. The overall-level metrics are further derived from the high- level metrics to obtain the overall engagement information. These three classes of metrics are adapted from several of those widely used engagement metrics for webpages discussed earlier in Sect. 2. Before we describe the proposed metrics, we first introduce a case study to which the metrics will be applied as an example.

4.1 Case Study

The example in Fig. 1 is used as a case study and revised with some information added as shown in Fig. 7. There are five CC or content in a container. c1, c2, and c3 are on page 1. At time = 1, c4 on page 2 has the same content (i.e. a music video A) as c2 on page 1. When time = 2, the container of c4 has its content changed to a music video B, and that location is then identified as c5. Table 4 summarizes the web content and container of each CC in the case study.

4.2 Low-Level User Engagement Metrics

Low-level user engagement metrics measure usage data directly from three sources, i.e. webpages, web contents, and containers. Since a content and a container share the same location on a webpage at the same time, we use the same low-level metrics for these two sources. Each metric is listed with its definition in Table 5.

Note that the metrics for webpage sources are taken from existing metrics but we enhance by proposing additional metrics for web contents and containers. Using the low-level metrics, we obtain the measurements of the case study in Tables 6 and 7.

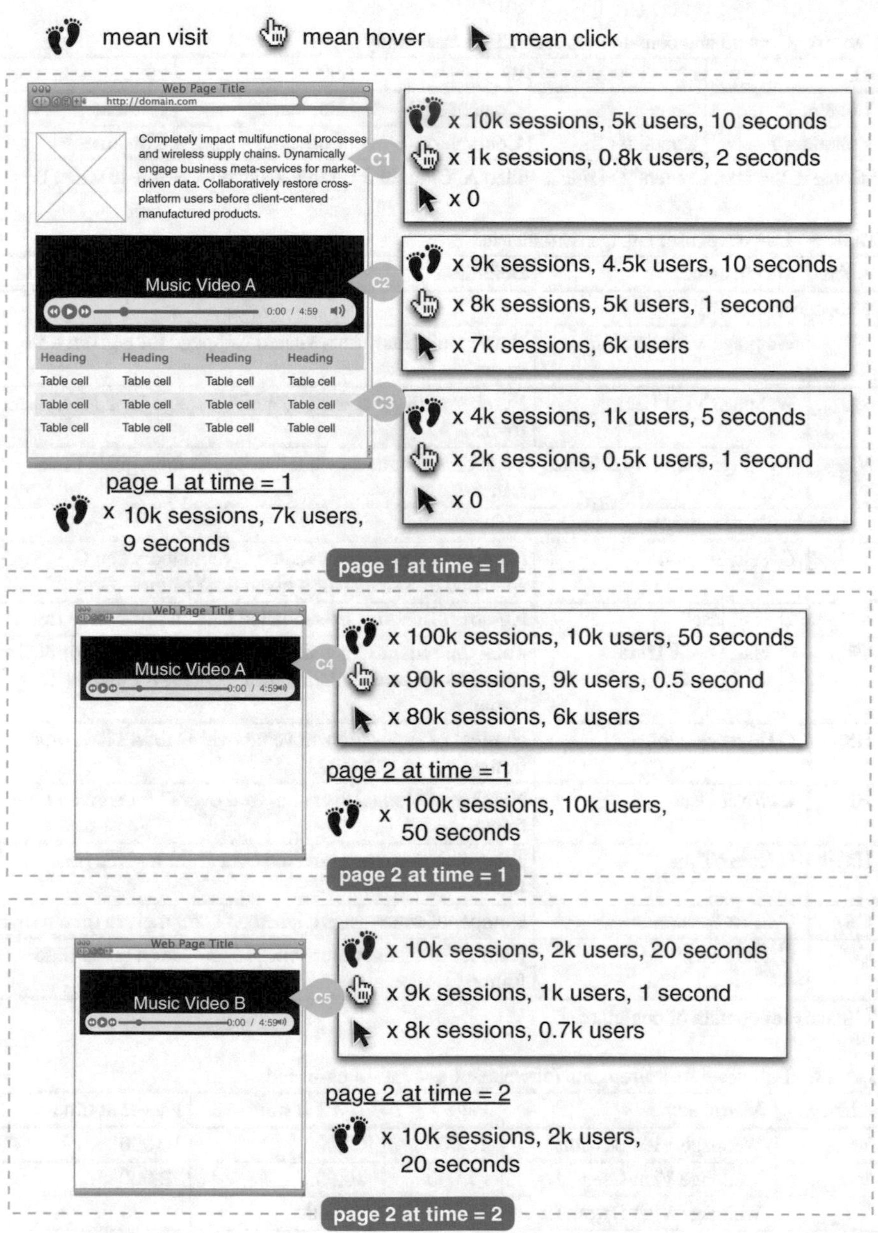

Fig. 7 Case study

Table 4 Content and container of each CC in case study

c1	c2	c3	c4	c5
Content 1[a]	Content 2[a]	Content 3[a]	Content 2	Content 4[a]
Container 1	Container 2	Container 3	Container 4	Container 4

[a] Content 1 = text, Content 2 = music video A, Content 3 = table, Content 4 = music video B

Table 5 Low-level user engagement metrics

Abbr.	Metric name	Definition
Webpage sources		
VS_{page}	Webpage Visit Session	Number of times users visited webpage for a given time frame
VU_{page}	Webpage Visit User	Number of distinct users visited webpage for a given time frame
VT_{page}	Webpage Visit Dwell Time	Total time that users visited webpage for a given time frame
Content/container sources[a]		
VS_C	C Visit Session	Number of times users visited C (counted when C appeared on screen) for a given time frame
VU_C	C Visit User	Number of distinct users visited C for a given time frame
VT_C	C Visit Dwell Time	Total time that users visited C (counted from the time C appeared on screen until C disappeared) for a given time frame
HS_C	C Hover Session	Number of times users hovered over C for a given time frame
HU_C	C Hover User	Number of distinct users hovered over C for a given time frame
HT_C	C Hover Time	Total time that users hovered over C for a given time frame
CS_C	C Click Session	Number of times users clicked on C for a given time frame
CU_C	C Click User	Number of distinct users clicked on C for a given time frame

[a]C stands for content or container

Table 6 Low-level measurements for webpage sources of case study

Abbr.	Metric name	Page 1	Page 2 at time = 1	Page 2 at time = 2
VS_{page}	Webpage Visit Session	10,000	100,000	10,000
VU_{page}	Webpage Visit User	7,000	10,000	2,000
VT_{page}	Webpage Visit Dwell Time	9	50	20

Table 7 Low-level measurements for content and container sources of case study

Abbr.	Metric name	c1	c2	c3	c4	c5
VS_C	C Visit Session	10,000	9,000	4,000	100,000	10,000
VU_C	C Visit User	5,000	4,500	1,000	10,000	2,000
VT_C	C Visit Dwell Time (seconds)	10	10	5	50	20
HS_C	C Hover Session	10,000	8,000	4,000	90,000	9,000
HU_C	C Hover User	800	5,000	500	9,000	1,000
HT_C	C Hover Time (seconds)	2	1	1	0.5	1
CS_C	C Click Session	0	7,000	0	80,000	8,000
CU_C	C Click User	0	6,000	0	6,000	700

4.3 High-Level User Engagement Metrics

High-level user engagement metrics are derived from low- level metrics and categorized by user behavior approaches, i.e. visit, click, and hover. Each metric is listed in Table 8 together with its definition. Table 9 shows the calculation results for the case study.

Table 8 High-level user engagement metrics

Abbr.	Metric name	Definition
Visit		
VSR_C	C Visit Session Rate	C Visit Session/C's Webpage Visit Session (VS_C/VS_{page})
VUR_C	C Visit User Rate	C Visit User/C's Webpage Visit User (VU_C/VU_{page})
VTR_C	C Visit Dwell Time Rate	C Visit Dwell Time /C's Webpage Visit Dwell Time (VT_C/VT_{page})
RR_C	Return Rate on C	C Visit User/Webpage Visit Session (VU_C/VS_{page})
Click		
SCR_C	C Session Clickthrough Rate	C Click Session/C's Webpage Visit Session (CS_C/VS_{page})
UCR_C	C User Clickthrough Rate	C Click User/C's Webpage Visit User (CU_C/VU_{page})
Hover		
SHR_C	C Session Hover Rate	C Hover Session/C's Webpage Visit Session (HS_C/VS_{page})
UHR_C	C User Hover Rate	C Hover User/C's Webpage Visit User (HS_C/VU_{page})
HTR_C	C Hover Time Rate	C Hover Time/C's Webpage Visit Dwell Time (HT_C/VT_{page})

High-level metrics can be used to determine user engagement with a web content in a container. However, to determine the overall user engagement with a web content that may appear in several containers as well as the overall user engagement with a container that may contain several contents over time, we need additional overall-level metrics.

4.4 Overall-Level User Engagement Metrics for Web Content

As depicted in Fig. 4, a web content may be contained in several containers across different webpages. To determine user engagement with this particular web content, we calculate an average engagement values of that web content over all containers in which it is contained:

$$Engagement_{content} = \frac{\sum_{i=1}^{n} Engagement_{content,i}}{n} \tag{1}$$

where $Engagement_{content}$ = overall user engagement with a web content at a particular time

n = number of containers in which that web content is contained, and

$Engagement_{content,i}$ = user engagement with that web content in the container i (obtained by using high-level metrics).

In the case study, c2 and c4 have the same content 2 (music video A). Using the engagement measurements for c2 and c4 in Table 9 as $Engagement_{content,i}$, we calculate the overall user engagement with the music video A at a particular time. In Table 10, the column "Content 2 at time = 1" lists the overall user engagement measurements for the music video A at time = 1. For example, $VSR_{content2} = (0.90 + 1.0)/2 = 0.95$ at time = 1.

Table 9 High-level measurements for content and container sources of case study

Abbr.	Metric name	c1	c2	c3	c4	c5
VSR_C	C Visit Session Rate	1.00	0.90	0.40	1.00	1.00
VUR_C	C Visit User Rate	0.71	0.64	0.14	1.00	1.00
VTR_C	C Visit Dwell Time Rate	1.11	1.11	0.56	1.00	0.40
RR_C	Return Rate on C	0.50	0.45	0.10	0.10	0.20
SCR_C	C Session Clickthrough Rate	0.00	0.70	0.00	0.80	0.80
UCR_C	C User Clickthrough Rate	0.00	0.86	0.00	0.60	0.35
SHR_C	C Session Hover Rate	0.00	0.70	0.00	0.80	0.80
UHR_C	C User Hover Rate	0.11	0.71	0.07	0.90	0.50
HTR_C	C Hover Time Rate	0.22	0.11	0.11	0.01	0.05

Table 10 Overall-level measurement of case study

Abbr.	Content				Container from time = 1 to time = 2			
	Time = 1		Time = 2					
	1	2	3	4	1	2	3	4
VSR_C	1.00	0.95	0.40	1.00	1.00	0.90	0.40	1.00
VUR_C	0.71	0.82	0.14	1.00	0.71	0.64	0.14	1.00
VTR_C	1.11	1.06	0.56	1.00	1.11	1.11	0.56	0.70
RR_C	0.50	0.28	0.10	0.10	0.50	0.45	0.10	0.15
SCR_C	0.00	0.75	0.00	0.80	0.00	0.70	0.00	0.80
UCR_C	0.00	0.73	0.00	0.60	0.00	0.86	0.00	0.48
SHR_C	0.00	0.75	0.00	0.80	0.00	0.70	0.00	0.80
UHR_C	0.11	0.81	0.07	0.90	0.11	0.71	0.07	0.70
HTR_C	0.22	0.06	0.11	0.01	0.22	0.11	0.11	0.03

4.5 *Overall-Level User Engagement Metrics for Web Container*

As depicted in Fig. 5, a container may contain several web contents at different time. It is useful to get an insight into how a particular container is engaged in order to design a container better or select suitable content for a container. To determine user engagement with a particular container over time, we calculate an average engagement values of that container over all contents contained in it:

$$Engagement_{container} = \frac{\sum_{i=1}^{t} Engagement_{container,i}}{t} \tag{2}$$

where $Engagement_{container}$ = overall user engagement with a web container over a range of time

t = number of times that user engagement with that container is determined (at a regular interval) over the time range, and

$Engagement_{container,i}$ = user engagement with that container measured at time i (obtained by using high-level metrics).

In the case study, c4 and c5 refer to the same container 4 (on webpage 2) with different contents (music video A and B) at different time. Using the engagement measurements for c4 and c5 in Table 9 as $Engagement_{container,i}$, we calculate the overall user engagement with this container 4 over a range of time. In Table 10, the column Container 4 from time = 1 to time = 2 lists the overall user engagement measurements for this container 4. For example, $VTR_{container4} = (1.0 + 0.4)/2 = 0.7$ over that time range.

5 Web-Contents-Based Analytics Tool

We develop a web-contents-based analytics tool that collects user interaction data, determine user engagements using the three classes of metrics, and visualize engagement information. The tool is implemented in PHP, with MySQL database. To install the tool on a website, a web developer has to register the website and generates jQuery code to put on the website. Once the code is installed, the tool will automatically collect necessary data, send to the server to process, and visualize the analytical results. The tool provides a dashboard by which the web developer can select particular web contents or containers (based on how the developers design the pages) and the engagement metrics that are of interest. Figure 8 shows an example of a graph displaying the UCR_C (User Clickthrough Rate) for the content 4 (music video B) at different time.

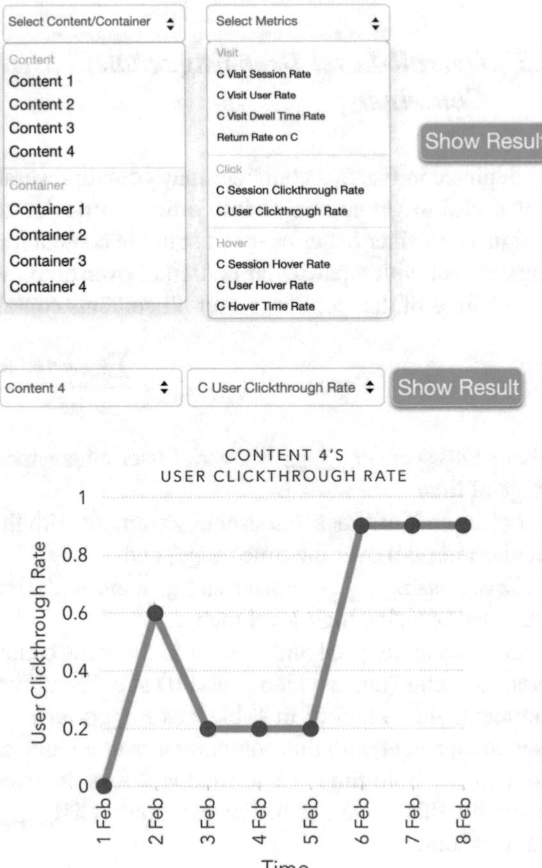

Fig. 8 Dashboard and visualization

6 Conclusion

The paper proposes an enhancement to existing web user engagement metrics by introducing additional engagement metrics that can take into account engagement with particular contents and areas on webpages. The analytical result is aimed to give an insight into user behavior and to help determine a way to improve website design by placing the right web contents at the right location. We are conducting a test on a commercial website that has installed the web-contents-based analytics tool in order to see if the site can benefit from the analytical results and improve their website design. Other future work includes improving the visualization of the tool and extending the set of engagement metrics.

References

1. Attfield, S., Kazai, G., Lalmas, M. et al.: Towards a science of user engagement (Position Paper). In: WSDM Workshop on User Modelling for Web Applications. http://www.dcs.gla.ac.uk/~mounia/Papers/engagement.pdf (2011). Accessed 9 April 2016
2. Datanyze: Analytics market share in the Alexa top 1M. https://www.datanyze.com/market-share/Analytics/ (2016). Accessed 9 April 2016
3. Google Analytics: Brian Gavin Diamonds sees 60 % increase in customer checkout with Google enhanced ecommerce. https://static.googleuserconent.com/media/www.google.com/en//intl/es%20_ALL/analytics/customers/pdfs/brian-gavin.pdf (2016). Accessed 9 April 2016
4. Google Analytics: With help from Periscopix, WBC employs advanced segments in Google Analytics to boost its e-commerce conversion rate by more than 12 %. https://static.googleusercontent.com/media/www.google.com/en//intl/es%20_ALL/analytics/customers/pdfs/wbc.pdf (2016). Accessed 9 April 2016
5. Lehmann, J., Lalmas, M., Baeza-Yates, R. et al.: Networked user engagement. In: Proceedings of the 1st Workshop on User Engagement Optimization (UEO'13) (2013)
6. Lehmann, J., Lalmas, M., Yom-Tov, E. et al.: Models of user engagement. In: Proceedings of the 20th International Conference on User Modeling, Adaptation, and Personalization (UMAP'12) (2012)
7. Lalmas, M., O'Brien, H., Yom-Tov, E.: Measuring User Engagement. In: Marchionini, G. (ed.) Synthesis Lectures on Information Concepts, Retrieval and Services. Morgan & Claypool Publishers (2015)
8. Huang, J., White, R., Dumais, S.: No clicks, no problem: using cursor movements to understand and improve search. In: Proceedings of the SIGCHI Conference on Human Factors in Computing Systems (CHI'11) (2011)
9. Google Analytics: Google Analytics Solutions. https://www.google.com/intl/en/analytics/#?modal_active=none (2016). Accessed 9 April 2016

Business Process Verification and Restructuring LTL Formula Based on Machine Learning Approach

Hiroki Horita, Hideaki Hirayama, Takeo Hayase, Yasuyuki Tahara and Akihiko Ohsuga

Abstract It is important to deal with rapidly changing environments (regulations, customer behavior change, and process improvement etc.) to keep achieving business goals. Therefore, verification for business process in various phases are needed to make sure of goal achievements. LTL (Linear Temporal Logic) verification is an important method for checking a specific property to be satisfied with business processes, but correctly writing formal language like LTL is difficult. Lacks of domain knowledge and knowledge of mathematical logics have bad influence on writing LTL formulas. In this paper, we use LTL verification and prediction based on decision tree learning for verification of specific properties. Furthermore, we helps writing properly LTL formula for representing the correct desirable property using decision tree constrction. We conducted a case study for evaluations.

1 Introduction

In recent years, organizations should deal with rapidly changing environments (regulations, customer behavior change, and process improvement etc.). Process-aware information systems (PAIS) [5] is an important means for these environmental

H. Horita (✉) · H. Hirayama · Y. Tahara · A. Ohsuga
The University of Electro-Communications, 1-5-1 Chofugaoka, Chofu,
Tokyo 182-8585, Japan
e-mail: h-horita@ohsuga.is.uec.ac.jp

H. Hirayama
e-mail: hirayama968@ybb.ne.jp

Y. Tahara
e-mail: tahara@is.uec.ac.jp

A. Ohsuga
e-mail: ohsuga@is.uec.ac.jp

T. Hayase
Toshiba Corporation, 1-1, Shibaura 1-chome, Minato-ku, Tokyo, Japan
e-mail: takeo.hayase@toshiba.co.jp

R. Lee (ed.), *Computer and Information Science*,
Studies in Computational Intelligence 656, DOI 10.1007/978-3-319-40171-3_7

changes to achieve business goals and filling gaps between business process and information system. Therefore, it is needed to verify business process for checking whether business goals are achieved or not. Verification of business processes are conducted in various phases (design time, runtime and backward verification). This research conducts on backward verification, and it is conducted on process mining [17] area. There are many research efforts in this area (e.g. process discovery and conformance checking). These methods can discover as-is model and analyze differences between observed behavior (event logs) and modeled behavior (process models). These methods can enhance effective environmental changes focusing attention on the whole of business process. On the one hand, LTL checker [16] can verify whether logs can satisfy desirable property or not using formal language like LTL. It is focussing attention on verifications of a local part of a business process.

In this paper, we propose a effective verification method focussing attention on verification of a local part of a business process. As an existing work, LTL checker can classify true traces and false traces for input properties. However, writing a formal language like LTL correctly is difficult. If we cannot write correct property using formal language, results like false positives and false negatives are raised which are undesirable. Therefore, in this paper, we verify whether each log has desirable property or not using LTL checker and predict false positives and false negatives based on decision tree learning. Furthermore, the decision tree constrcution is used for restructuring the logical formula. That can help users who do not have the domain knowledge and mathematical logic knowledge.

We conducted evaluation using logs of phone repair processes. The results showed that our proposed method can detect false positives and false negatives, and can predict goal achievements. These results can be used to describe a more precise LTL formula.

This paper is organized as below. Section 2 explains the background and related works. Section 3 explains proposed method. Section 4 explains evaluation results. Section 5 explains discussions. Finally, Sect. 6 explains conclusions and future works.

2 Background and Related Works

In this section, we explain backgrounds and related works.

2.1 Verfication of Properties in Business Process Managment

For developments and improvements of PAIS, it is needed to check business process adequateness of envisioned operational environments of the former. The adequate business process is a prerequisite of a PAIS operation. However, designing business process that can achieve business goals is difficult for many reasons. Business process verifications are conducted in some phases. The first is the verification of designs. For developments and improvements of PAIS, it is needed to check that designed business process models can satisfy specifications. The second comes runtime verification. It supports detecting inadequate event sequences and data in PAIS operation at runtime. The third comes backward verification. It analyzes past event logs for checking the adequacy. Backward verification has two areas. The first method is verification of differences between observed behaviors (event logs) and modeled behaviors (process models). It is called conformance checking.

Rozinat et al. proposed token replay conformance checking method [14]. It measures the fitness of the process model and event logs. Leoni et al. proposed a multi perspective conformance checking method [3]. This technique deals with control flow, data and resource for the conformance checking. Molka et al. proposed a conformance checking method for BPMN-based process models [10]. Their approach can check conformance between BPMN models and event logs on a local level (individual parts of the process and individual events) and global conformance measure (a trace level). It analyzes deviation parts and distance (it is represented as fitness metrics) between logs and models. Prior and the latter has different purposes; therefore we should use these methods depending on the situation.

The second is verifying specific properties of business processes. We can verify if specific formula is true or false in each trace using formal languages. Business goals and rules can be represented as a formal language like Event Calculus [11] and linear temporal logic (LTL) [13]. We used LTL checker [16] for parts of business process verification and explain LTL.

Linear temporal logic is possible to specify constraints relating with temporally changing property. A formal verification method using linear temporal logic can automatically and exhaustively verify whether the target of a verification satisfies desired property. It is used for recent complex information systems and business processes. Linear temporal logic provides the classical logic operators ($\neg$, $\wedge$, $\vee$, $\Rightarrow$ and $\Leftrightarrow$), and uses several temporal operators ($\circ$(nexttime), $\diamond$(eventually), [](always), U(until)) that can be used to specify constraints over the sequencing of workflow tasks [5]. In process mining area, [16] is a verification method using linear temporal logic called LTL checker. It is a plug-in used on ProM.[1] ProM is an open source framework for process mining. We use the LTL checker for business process verification, so we explain this method using simple examples. Table 1 represents simple logs. Each log has an ID and sequence of conducted events (it is called traces). Each event is represented by a single alphabet (A, B, C, D, E). For example, in ID 1, event A, B, C, D are conducted in turn. Table 2 represents constraints of workflows. Each constraint is described using natural languages and formally defined by linear temporal logic. The LTL checker can verify that traces in Table 1 satisfy constraints in Table 2. Both case 1 and 2 can satisfy all constraints, but the case 3 violates "D or E should eventually occur in a trace". In case 3, both event D and E are not conducted. At lowest, only event D and E should be conducted for satisfying "D or E should eventually occur in a trace".

[1]http://www.processmining.org/prom/start.

Table 1 Simple example of event logs

ID of traces	Traces
1	ABCD
2	ABED
3	ABC

Table 2 Constraint examples

Constraint	Formal constraint
After every occurrence of A eventually B should occur in a trace	$A \Rightarrow \diamond B$
D or E should eventually occur in a trace	$\diamond(D \vee E)$

Medeiros et al. proposed semantic LTL checker [4]. It extended the LTL checker and can verify business process logs using ontology that can represent the element equal semantically.

2.2 Decision Tree Learning

Decision tree learning is a supervised learning technique aiming at the classification of instances based on predictor variables. There is one categorical response variable labeling the data and the result is arranged in the form of a tree [17]. Decision trees can represent decision rules by learning data and can represent decision rules graphically as a tree. A decision tree is constructed from the obtained training data. In this paper, instances are traces, and a response variable is whether desirable property is satisfied or not. For using structured data like event traces, how to decide the feature set is not clear. We use 2-grams of each event trace as feature sets. In this way, we can represent event order of each trace.

Maggi et al. proposed a goal achievement prediction method using decision tree learning [8]. This method can predict goal achievement probability at runtime. Our paper conduct backward verification using decision tree and a restructuring logical formula method. It takes account situations that incorrect logical formulas are used for verification and supports users who do not have the domain knowledge and mathematical logic knowledge.

2.3 N-Grams in Business Process Management

N-gram represents sentence as adjacent n string. Especially, in case of "n = 1" is uni-gram, "n = 2" is bi-gram, "n = 3" is tri-gram. For example, "ABCDEF" is split to follow in case of 2-grams.

$ABCDEF \Rightarrow \{\text{“}AB\text{”}, \text{“}BC\text{”}, \text{“}CD\text{”}, \text{“}DE\text{”}, \text{“}EF\text{”}\}$

In this way, string sequences are split to a partial structure of it by n-grams.

N-grams are widely used in natural language processing and information retrieval [1]. These ideas are used in also business process management. Mahleko et al. proposed a indexing method for effective matching of web service infrastructure by conducting transformation from business processes to n-grams [9]. Wombacher proposed a similarity measuring method of workflows for service discovery representing a business process structure as n-grams [18].

Our method uses n-gram as features representing event orders in each trace. It can help prediction, and restructure a logical formula.

3 Proposed Method

In this section, we explain our proposed method. Our method which is more accurate for prediction of logical formula verification and a method which is logical formula restructuring. It detects false positives and false negatives at the verification for compensating inadequate logical formulas. Existing approaches are inadequacy when user does not have a adequate domain knowledge and knowledge of mathematical logics. Our method can deal with this problem focussing on a part of event trace structure (n-grams) and predcition using decision trees.

We explain false detection and oversight detection using Fig. 1 in this paper. Figure 1 represents detected traces and undetected when a logical formula verification is conducted. If traces satisfy the specific property, these traces are classified into goal achieving traces (green). Red rectangle represents desirable property satisfying traces on an equality with green rectangle, but the specific property is not true, because the logical formula does not accurately represent the desirable property. This phenomenon is called oversight detection in this paper. On the other front, there are incorrect traces judged as satisfying the specific property. It is inadequate classification, because these traces cannot satisfy the desirable property. It is classified into the purple rectangle. This phenomenon is called false detection. Both phenomena

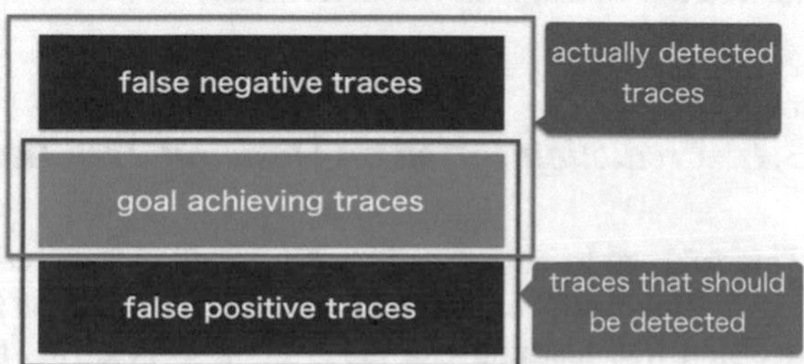

Fig. 1 False positive traces and false negative traces

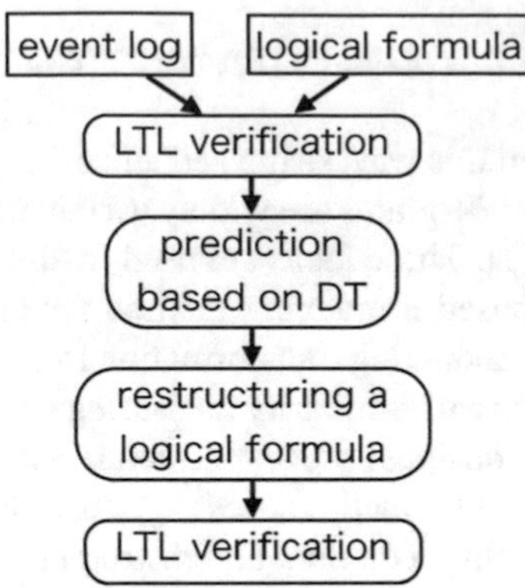

Fig. 2 Overview of our proposed method

are not desirable, therefore they should be decreased. What is desirable truly is the discovery of traces satisfying the desirable property. Our method can detect both of these undesirable things, and help constructing a more accurate desirable property.

Figure 2 represents an overview of our proposed method. By this process, traces are detected as satisfying desirable property by prediction based on decision tree and reconstructuring logical formula method. In this process at first, consistency between a logical formula represented as LTL and event logs is verified. If input traces satisfy an inputed logical formula, these traces are classified into true traces and do not satisfy the logical formula, the traces are classified into false traces. We use LTL checker [16] for the verification in this part. Second, false detection and oversight detection are conducted. For true traces, we conduct false detection because we should remove these traces that have the possibility of essentially, the non-desirable property. For false traces, we conduct oversight detection, because we should discover truly desirable property satisfied traces. Using these results, we can know the number of the truly desirable property satisfied traces as compared to by the results by using logical formula verification only. False detection and oversight detection are conducted by prediction using a decision tree. We will explain it in Sect. 3.1. Third, we restructure a logical formula based on a construction of the decision tree. Decision trees have information relating conditions for the desirable formula. The restructured logical formula is more correct than the initial formula. At the end, LTL verification is conducted using the restructured logical formula. The results are higher accuracy rate than results of using the initial formula.

3.1 Prediction Based on Desicion Tree Learning

This subsection corresponds with "prediction based on DT (Decision Tree)" in Fig. 2. The prediction is conducted based on decision tree learning. Prediction using decision tree learning can discover false positives and false negatives. Decision tree learning is a supervised learning technique, therefore it uses training data for constructing a prediction model. Each part of true trace group and false trace group are training data. A decision tree is constructed using these training data based on Classification And Regression Tree (CART) algorithm [7].

Training data are constructed from parts of each trace as n-grams. Traces are structured data, so there are not vector data. Many machine learning algorithms use vector data for learning. Therefore, we transform event logs represented as structured data to vector data focussing on n-grams of event traces. Each trace is transformed to some n-grams. Sets of each number of n-grams in each trace are feature sets. Each trace has an answer label. Whether it represents the trace is true or false if the correct logical formula are used for the verification. Therefore, each trace has features about events ordering. Each event name is transformed to an alphabet. For example, when a trace has a feature AB, it represents that If event A is conducted, event B is conducted just after it. A trace have normally a number of n-gram relations, so a set of n-grams features represents event order of each trace. Generated Decision trees represent conditional branching that represents whether each trace is true or false.

Table 3 is an example representing feature vectors, and each trace have a label representing the class (true or false). Each row is trace ID, event sequence, a 2-gram feature and a label. For example, an event sequence of trace 1 is "ABBCAB". Each alphabet represents an event which is recorded in business process logs. Event sequences represent an order of events. For learning using a decision tree, feature sets are 2-gram columns in a Table 3. In each 2-gram columns, a value is counted in each trace. Trace 1 has some 2-grams (AB, BB, BC, CA, AB). Only AB is counted twice and recorded as 2. Another are 1. The feature set of logs is all 2-grams of the logs. Not existing 2-gram in each trace are recorded as 0. Using these information, a decision tree is learned by one logical formula verification.

A learned decision tree can predict the label (true or false) of each trace. Figure 3 is an example of a decision tree. The tree is split twice. At root node, the split condition is DE $\leq$ 0.5. It represents if DE is in the trace, it classified into False, if not, into True. In the same way, the under node split is conducted. A value is classified results in the node. The left side represents true and the right side represents false. A class is prediction results. Both numbers of class label 1 and 0 are 50 $(50, 42 + 8)$.

Table 3 Example of rraces having 2-grams as features

Trace ID	Event sequence	AB	BB	BC	CA	...	Label
1	ABBCAB	2	1	1	1	...	True
2	ABCD	1	0	1	0	...	False
3	ABBA	1	1	0	0	...	False

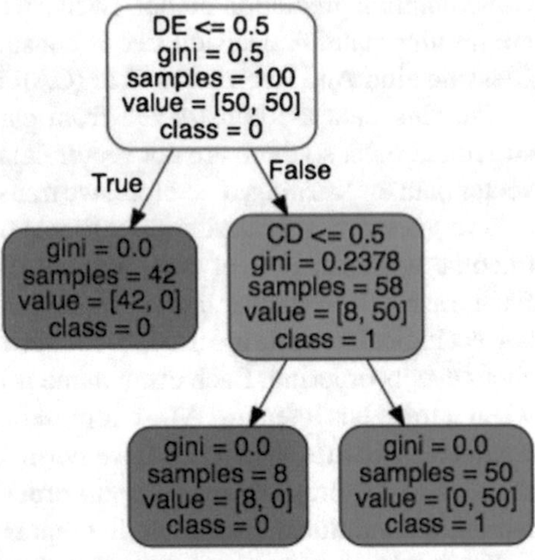

Fig. 3 A decision tree (2-grams of Event Traces)

3.2 *Logical Formula Restructuring Based on Decision Tree Construction*

The decision tree constructed in Sect. 3.1 are used for restructuring a logical formula. That is conducted for restructuring logical formula which is not reflected correct intentions to more correct logical formula. That contributes to more accurate verification.

Next, we explain a method to transform a decision tree to a logical formula. We use decision tree which is split by 2-grams in incremental steps. In that decision tree, we should pay attention to leaf node which is having class label 1 and all conditional branching from the root node to the leaf node, and construct a conjunction of all conditional branching. Conditional branching is represented as AG $\geq$ 0.5, it represents event A are conducted and after that, event G are conducted. If it is true, logical formula is $\diamond(A \wedge \circ G)$, and if not, it is $\neg\diamond(A \wedge \circ G)$. In addition, all passes are disjuncted as $\vee apass$. It implies choice is many for class label 1. In the results, the formula can cover all conditions for class label 1. That is to say, in a decision tree, only parts of class label 1 are transformed to the logical formula. General form of logical formula constructed by decision tree is follows.

$\vee${$\wedge$a pass from the root node to a leaf node having “class label = 1”}

Conjuncting upper logical formula and initial logical formula, more correct formula are constructed. General form of this is follows.

(a initial logical formula$\wedge\vee${$\wedge$a pass from the root node to a leaf node having “class label = 1”})

Using this formula for verification, more higher rate results are earned. This phase is second LTL verification in Fig. 2.

4 Validation

In this section, we evaluate the effectiveness of our proposed method by conducting a case study and comparing our proposed method with other methods. Datasets and its logical formulas for evaluation are explained in Sect. 4.1. In Sect. 4.2, we construct a decision tree and restructuring logical formula using its tree. In Sect. 4.3, we compare verification, prediction, verification using reconstructed logical formula and another method [6].

4.1 Datasets and Logical Formulas

We have evaluated our approach on an event log which is taken from a phone repair process and is publicly available and used in some researches for evaluations. The log contains 11855 events from 12 different events in 1104 cases, each case representing a phone terminal repair process (register, analyze defect, repair, test repair, archive and etc.).

The log deta format is XES,[2] which is xml-based standard for event logs and used in tools like ProM. Each trace describes a sequential list of events corresponding to a particular case. The log, its traces, and its events may have any number of attributes [17]. Attributes are standard (case id, time and etc.) or domain specific (phone type, defect type and etc.).

We used 3 pairs of correct logical formula and incorrect logical formula relating a phone repair process. Following formulas are part of them. Correct represents desirable property, and incorrect represents a logical formula which does not reflect user's intention adequately. Such a situation happens for reasons like inadequate domain knowledge and knowledge of mathematical logic. Both formulas are similar, but these have different meaning. Therefore, verification results are different. Correct formula represents if activity == "Repair (Complex) start" (A) conducted eventually event activity == "Repair (Complex) complete" (B) is conducted, after that, eventually event activity == "Inform User" (C) are conducted, so it represents the permissible order of events are only one. On another front, incorrect formula represents that events A, B and C eventually conducted. This formula does not specify any constraints on the order of these 3 events (ABC, BAC, CBA and etc. are accepted). These logical formulas are checked by LTL checker and this is a LTL verification part of Fig. 2.

[2]http://www.xes-standard.org/.

$$
\begin{aligned}
correct &: \Diamond((activity == Repair(Complex)start \\
&\quad \wedge \Diamond ((activity == Repair(Complex)complete \\
&\quad \wedge \Diamond\, activity == InformUser)))) \\
incorrect &: ((\Diamond activity == Repair(Complex)start \\
&\quad \wedge \Diamond\, activity == Repair(Complex)complete) \\
&\quad \wedge \Diamond\, activity == InformUser)
\end{aligned}
$$

4.2 Constructing Decision Tree and Logical Formula Improvements Results

For constructing a decision tree, we have split log to the first part as training data and the second part as test data. Training data are chosen at random from each true and false group, and have a label representing desirable property is true or false. Using these training data, a decision tree is constructed by CART algorithm [7] using scikit-learn [12]. We conducted 10-fold cross validation. Figure 3 is constructed from training data. Section 3.1 explained about this decision tree.

Next, we explain the results of reconstructing logical formula constructed from the decision tree explained in Sect. 4.1. First, logical formula generated from the decision tree is follows.

$$
\begin{aligned}
&(\Diamond((activity == Repair(Complex)start \\
&\wedge\circ(activity == Repair(Complex)complete))) \\
&\wedge \Diamond ((activity == AnalyzeDefectcomplete \\
&\wedge\circ(activity == Repair(Complex)start))
\end{aligned}
$$

The conjunction of the former formula and initial formula is follows.

$$
\begin{aligned}
&(((\Diamond(activity == Repair(Complex)start) \\
&\wedge \Diamond (activity == Repair(Complex)complete)) \\
&\wedge \Diamond (activity == InformUser)) \\
&\wedge(\Diamond((activity == Repair(Complex)start \\
&\wedge\circ(activity == Repair(Complex)complete))) \\
&\wedge \Diamond ((activity == AnalyzeDefectcomplete \\
&\wedge\circ(activity == Repair(Complex)start)))))
\end{aligned}
$$

In this way, initial formula is reconstructed to more correct formula. In next section, adequacy of the formula is evaluated.

4.3 Prediction Results Based on Decision Tree

In this section, we compare each part of results in our proposed process explained in Fig. 2 and compare with another method [6].

Tables 4, 5 are results of verification using initial formula. These results showed many misclassifications. Tables 6, 7 are prediction results using the decision tree and Tables 8, 9 are results of verification using the restructured formula. Both Tables 6, 7, 8, and 9 showed good results than Tables 4, 5. Figure 4 is a comparison of results of initial formula verification, decision tree prediction, restructured formula verification and another approach [6]. The other approach uses decision tree prediction, but feature set for learning is different from our proposed approach. The feature set is each event execution (each event is conducted or not). The results are a comparison of accuracy rates (accuracy rate = (TP + TN)/(FP + FN + TP + TN)). The results showed accuracy rate of our approach (decision tree prediction using 2-gram and restructuring logical formula) is higher than initial formula verification and another approach. In this way, prediction based on decision tree using 2-gram of event traces as feature set and logical formula restructuring is effective.

Table 4 Results of initial verification (true)

		Actual	
		T	F
Classified	T	243	0
	F	414	0

Table 5 Results of initial verification (false)

		Actual	
		T	F
Classified	T	0	0
	F	0	447

Table 6 Results of prediction (true)

		Actual	
		T	F
Classified	T	224	19
	F	2	412

Table 7 Results of prediction (false)

		Actual	
		T	F
Classified	T	0	0
	F	2	445

Table 8 Results of Verification Using Restructured Logical Formula (true)

		Actual	
		T	F
Classified	T	240	0
	F	16	0

Table 9 Results of Verification Using Restructured Logical Formula (true)

		Actual	
		T	F
Classified	T	0	3
	F	0	845

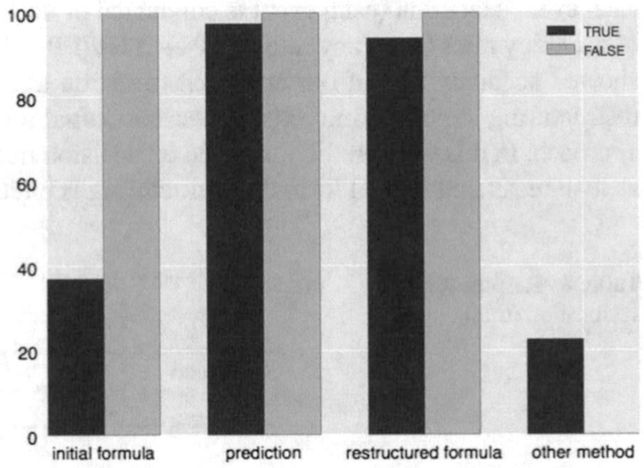

Fig. 4 Results of comparison in each process step and other method

5 Discussion

In this section, we discuss the validity of our approach and the limitation.

Experimental results showed that our approach can verify and predict about desirable property and a restructured logical formula. Accuracy rates are higher than initial verification and another method. This helps users who have not the adequate domain knowledge and mathematical logics knowledge.

We used a decision tree for prediction and restructured logical formula. Other classifiers like a support vector machine and naive bayes also can predict. Especially, kernel method [15] and support vector machine is an important method for learning of structured data. Kernel method can predict without explicit feature sets using a kernel according to types of structured data. Figure 5 is comparison results of some classifiers. Accuracy rates of decision tree is higher than the ones of naive bayes and simmilar with SVM, but these comparison methods are inadequacy for restructured logical formula, because these two methods do not have structures like decision tree. We use a tree structure of decision tree. Therefore, Using a decision tree is effective for the prediction and the verification.

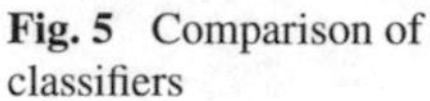

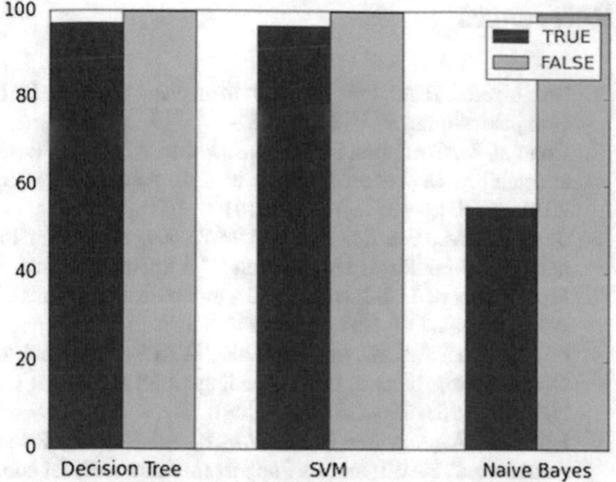

Fig. 5 Comparison of classifiers

We compared prediction results when using 2-grams, 3-grams and 4-grams of event traces as features. The results are similar high accuracy rate, therefore 3-grams and 4-grams can be used as well. However, when 3-grams and 4-grams are used, restructured logical formulas are more complex than when using 2-grams.

Verifications using restructured logical formulas are good results, but it is not most suitable for verification of desirable property. The formula is redundant and different from desirable property. A further good method for earning correct and short logical formula is a future work.

Our method has limitation relating with scope of application. We use only 2-grams of event logs as feature set. Therefore, we use information about event sequence, but event logs also have other elements like timestamp and resource data. Verification relating these elements property (e.g. Repairing of the telephone has to end in 1 day.) are not conducted effectively, but properties relating event sequences can be effectively conducted.

6 Conclusion

In this research, we proposed a prediction method using decision tree for compensating an incorrect logical formula. Moreover, our method helps to specify a more precise logical formula to be satisfied in business processes.

We compared our method to an other method and represented effectiveness of our method by conducting case studies. Our approach can predict true or false of logical formulas using features represented as 2-grams of event sequences. Furthermore, we represented that restructuring method of logical formula based on decision tree construction is effective. It helps users who have not domain knowledge and have not a detailed knowledge of mathematical logics.

Future works will be using our methods for runtime analysis [2]. Furthermore, using various information about timestamps and resource data is a future work.

References

1. Baeza-Yates, R.A.: Text retrieval: theory and practice. In: Proceedings of the 12th IFIP World Computer Congress 1992, pp. 465–476 (1992)
2. Cognini, R., Corradini, F., Gnesi, S., Polini, A., Re, B.: Research challenges in business process adaptability. In: Proceedings of the 29th Annual ACM Symposium on Applied Computing 2014, pp. 1049–1054. ACM (2014)
3. de Leoni, M., van der Aalst, W.M.P.: Aligning event logs and process models for multi-perspective conformance checking: An approach based on integer linear programming. In: Proceedings of 11th International Conference on Business Process Management 2013. LNCS, vol. 8094, pp. 113–129. Springer (2013)
4. de Medeiros, A.K.A., van der Aalst, W.M.P., Pedrinaci, C.: Semantic process mining tools: Core building blocks. In: Proceedings 16th European Conference on Information Systems 2008, pp. 1953–1964, AISeL (2008)
5. Hofstede, A., van der Aalst, W.M.P., Adams, M., Russell, N.: Modern Business Process Automation: YAWL and Its Support Environment, 1st edn. Springer, Berlin (2009)
6. Horita, H., Hirayama, H., Hayase, T., Tahara, Y., Ohsuga, A.: Goal achievement analysis based on LTL checking and decision tree for improvements of PAIS. In: Proceedings of 31th Annual ACM Symposium on Applied Computing 2016, pp. 1214–1216. ACM (2016)
7. Loh, W.Y.: Classification and regression trees. Data Min. Knowl. Disc. **1**(1), 1423 (2011)
8. Maggi, F.M., Di Francescomarino, C., Dumas, M., Ghidini, C.: Predictive monitoring of business processes. In: Proceedings of 26th International Conference on Advanced Information Systems Engineering 2014. LNCS, vol. 8484, pp. 457–472. Springer (2014)
9. Mahleko, B., Wombacher, A.: Indexing business processes based on annotated finite state automata. In: Proceedings of International Conference on Web Services, pp. 303–311. IEEE Computer Society (2006)
10. Molka, T., Redlich, D., Drobek, M., Caetano, A., Zeng, X.J., Gilani, W.: Conformance checking for BPMN-Based process models. In: Proceedings of 29th Annual ACM Symposium on Applied Computing 2014, pp. 1406–1413. ACM (2014)
11. Montali, M., Maggi, F.M., Chesani, F., Mello, P., van der Aalst, W.M.P.: Monitoring business constraints with the event calculus. ACM Trans. Intell. Syst. Technol. **5**(1), 17:1–17:30 (2014)
12. Pedregosa, et al.: Scikit-learn machine learning in python. J. Mach. Learn. Res. 12-2825-2830 (2011)
13. Pnueli, A.: The temporal logic of programs. In: Proceedings of the 18th Annual Symposium on Foundations of Computer Science 1977, pp. 46–57. IEEE Computer Society (1977)
14. Rozinat, A., van der Aalst, W.M.P.: Conformance checking of processes based on monitoring real behavior. Inf. Syst. **33**(1), 64–95 (2008)
15. Shawe Taylor, J., Cristianini, N.: Kernel Methods for Pattern Analysis. 1st edn., Cambridge University Press (2004)
16. van der Aalst, W.M.P., de Beer, H.T., van Dongen B.F.: Process mining and verification of properties: an approach based on temporal logic. In: Proceedings of On the Move to Meaningful Internet Systems 2005: CoopIS, DOA, and ODBASE. LNCS, vol. 3760, pp. 130–147. Springer (2005)
17. van der Aalst, W.M.P.: Process Mining: Discovery, Conformance and Enhancement of Business Processes, 1st edn. Springer, Berlin (2011)
18. Wombacher, A.: Evaluation of technical measures for workflow similarity based on a pilot study. In: Proceedings of on the Move to Meaningful Internet Systems 2006: CoopIS, DOA, GADA, and ODBASE. LNCS, vol. 4275, pp. 255–272. Springer (2006)

Development of a LMS with Dynamic Support Functions for Active Learning

Isao Kikukawa, Chise Aritomi and Youzou Miyadera

Abstract In this study, we propose a new type of Learning Management System (LMS), which gives teachers more freedom in planning active learning. We name it "Dynamic LMS (DLMS)". The DLMS generates user interface, in which learners access information necessary for activities, by interpreting a "Learning Design Object (LDO) package". It is based on a minimum expansion of the Instructional Management System-Learning Design (IMS-LD). It offers teachers a lot of choices. They can choose self-learning, a pair work or a group work, and even change roles in a group later. It is constructed from the "LD editor" and "LD player" that are compatible with the LDO package. These linked functions enable teachers to plan a variety of activities and, consequently, create more effective active learning in their classes.

Keywords LMS · Active learning · Course design · Learning design · IMS-LD

1 Introduction

In recent years, active learning has been popular in many classes all over the world. There are many kinds of activities such as self-learning, a pair work and a group work. Active learning is especially effective when learners engage in a few different kinds of activities, which we call "the flow of activities". Though LMS is supposed to assist it effectively, no single LMS in the previous studies enable teachers to plan

I. Kikukawa (✉) · C. Aritomi
Tokoha University, 325, Obuchi, Fuji-shi, Shizuoka 417-0801, Japan
e-mail: kikukawa@fj.tokoha-u.ac.jp

C. Aritomi
e-mail: aritomi@fj.tokoha-u.ac.jp

Y. Miyadera
Tokyo Gakugei University, 4-1-1, Nukuikita-machi, Koganei-shi,
Tokyo 184-8501, Japan
e-mail: miyadera@u-gakugei.ac.jp

R. Lee (ed.), *Computer and Information Science*,
Studies in Computational Intelligence 656, DOI 10.1007/978-3-319-40171-3_8

a variety of activities because each LMS is specified for one particular activity respectively.

This study, therefore, aims to develop a new LMS in order to support "the flow of activities". More concretely, it aims to develop a LMS with which teachers can plan a variety of activities. They first choose one activity out of self-learning, a pair work, and a group work. Choosing a group work, for example, they assign groups and later, can change learners' roles within a group. We name the whole procedure "dynamic planning of activities".

The IMS-LD [1] was used in order to support current LMS. It is based on the specification of the IMS-LD and a minimum expansion of the IMS-LD. This paper describes the design and development of our new LMS named DMLS, which enables teachers "dynamic planning of activities".

2 Dynamic Planning of Activities

2.1 Activities Required in Active Learning

In this section, activities required in active learning are discussed with reference to Fink's concept of "Creating Significant Learning Experiences [2]".

Fink identifies three components, "Information and Ideas", "Experiences" and "Reflecting (Reflective Dialogue)" as important elements in active learning. He also states that "an effective set of learning activities is to include activities from each of the three components", and these activities can be carried out by one person alone or with other persons.

Based on Fink's concept [2], the requirements for active learning can be summarized as follows:

- There are three important components for active learning, which are "Information and Ideas", "Experiences" and "Reflecting".
- An effective set of activities should contain all the three components as a whole.
- Activities can be a self-learning, a pair work or a group work.

2.2 The Outline of Learning to be Realized by the DLMS

In this section, we will outline what teachers can do using the DLMS we are presenting in the study.

Figure 1 shows a procedure of "dynamic planning of activities". It consists of three Learning Units. Each Learning Unit contains a collection of activities, which are associated with the roles. Each activity has built-in resources to be used as learning materials.

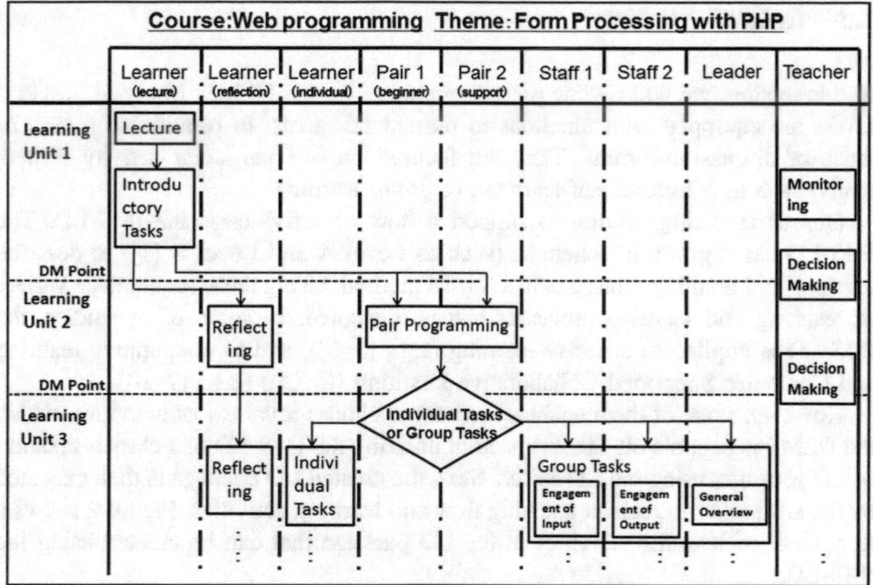

Fig. 1 Example of "dynamic planning of activities"

In the Learning Unit 1, the teacher introduces the thematic content to learners by lecture and they engage in introductory tasks. The teacher monitors them to see how they are doing and if they are done with the Unit 1. She/he then has a choice, choosing who and who will be partners, and which person will be "Pair 1" or "Pair 2". We call it a "Decision Making Point (DM Point)" in this study.

In the Learning Unit 2, they have two tasks at the same time. One is a programming task in a pair work, and the other is a reflection task in a self-learning. In Pair Programming, they have several tasks. Every time they finish a task, they must write down what they learned to keep a record in their "learning journal". An advantage of the DLMS is that they can engage in two or more activities in each unit, which we call "multi-roles".

When the Learning Unit 2 is finished, here comes the second DM Point. The teacher can choose either "Individual Tasks" or "Group Tasks". Based on her/his choice, they proceed to the Learning Unit 3 and engage in the activity. If she/he chooses a group task, each learner will be assigned to a role of either "Leader", "Staff 1" or "Staff 2". One of the advantages of the DLMS is that they can switch their roles within a group. In addition to a group work, they take a role of reflection in the Learning Unit 3 as well. When a group work is in process, they must also record what they did in the learning journal at the same time.

As shown above, the DLMS enables the teacher to plan more activities for learners, and thus offer them more effective active learning.

2.3 Related Studies

In this section, we will review preceding studies on the DLMS. Many of previous LMSs are equipped with functions to present resources, to provide mini-tests or those of discussion-forums. They are focused on one particular activity respectively, such as a lecture, self-learning, or group learning.

One of preceding studies to support a flow of activities is the IMS-LD. The IMS-LD has a group of schemata (such as Level A and Level B [1]) to describe teaching and learning processes in a formal method. Using these schemata, a variety of teaching and learning processes can be designed. In some other studies, the IMS-LD is applied to adaptive learning (e.g., [3–6]), and to cooperative learning and Computer Supported Collaborative Learning (CSCL) (e.g., [7–10].

However, none of them enables teachers to choose a variety of activities, which the DLMS is capable of. The first step of utilizing the IMS-LD in a class is creation of LD package, using the LD editor. Next, the created LD package is then executed by the LD player to realizer teaching flow and learning activities. Figure 2 shows a flow chart of learning activities in the LD package that can be created using the IMS-LD.

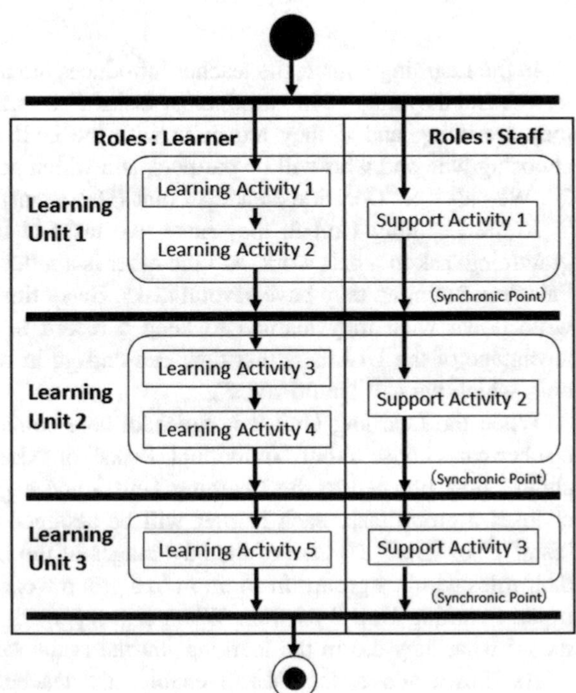

Fig. 2 Flow of learning activities using IMS-LD

Each "Learning Unit" is separated by a "Synchronic Point", which determines the timing to proceed to the next "Learning Unit". The "Learning Unit" is a combination of activities, which are associated with the roles of both learners and staffs. The activity contains resources to be used as learning materials. Besides, there are two roles. One is "Learner" and the other is "Staff". The activities associated with "Learner" are referred to as "Learning Activities" and those associated with "Staff" as "Support Activities". Though the "Learning Units" in the IMS-LD can be partially used as a constitutive element of a "flow of learning activities", it does not give teachers a freedom of choosing activities which the DLMS can offer, because the IMS-LD has three problems as follows:

- Problem 1: It does not allow description of information which is necessary to change learners' roles, move on to another activity, or assign two different tasks at the same time.
- Problem 2: It is not equipped with a function to assign who the leader in a group is, who Staff 1 is, or how many constitute a group.
- Problem 3: At the "Synchronic Point" it only allows teachers to decide when to move forward to next unit. They cannot change an activity to another.

Another study on the IMS-LD was developed by Dalziel, and named LAMS [11]. However, using LAMS, a teacher cannot give various roles to learners at one time.

It is, therefore, concluded that no LMSs nor IMS-LD in previous studies can offer learners as many activities, which the DLMS is aimed to achieve.

3 The Designing of the DLMS

3.1 Towards the Realization of the DLMS

In order to develop the DLMS which effectively supports active learning activities, we focus on minimum expansion on the IMS-LD in accordance with IMS-LD specification. It will be designed to have ability of "dynamic planning of activities", with which teachers will be able to choose such things as a pair work, a group work, changing roles within a group, or making new groups. Moreover, the DLMS will be able to generate "interface" on learners' screens, which has many useful functions for them to use during their activities. Advantages of expanding the IMS-LD include ability to use the IMS-LD schema groups which are already mentioned and ability to use existing tools (such as the LD engine [12]). These advantages can help focus on the content and work for the parts to be developed in the study, which will improve efficiency.

3.2 The LDO Package and the DLMS

In this section, we will explain how to expand the IMS-LD and an overview of LDO package, a new package obtained by expansion of the IMS-LD.

At first, we assume that a "flow of learning activities" is a "combination of various parts consisting of IMS-LD Level A and IMS-LD Level B elements". At the same time, the "IMS-LD parts" are defined as the "IMS-LDO". In addition, the Level A LDO is referred to as LDO_A and the Level B LDO is referred to as LDO_B. The "flow of learning activities" consisting of multiple LDOs is defined as the "LDO package". It consists of the following LDOs, which is expected to solve the problems listed in Sect. 2.3 successfully.

- LDO_A: LDOs is used to constitute the base of the "flow of learning activities" ([roles], [learning activity], [support activity] and so on which are used in the [Learning Unit]).
- LDO_B: LDOs is used to complete the "flow of learning activities" by adding to the base created in LDO_A and to assign learners to [learning activities], [roles] and [groups] which are defined in the next [Learning Unit] at the "DM Point" ([learning activity management]).

More detailed definitions of LDO_A and LDO_B are described in the following section.

Figure 3 is an overview of the DLMS. The DLMS will be equipped with LD editor and LD player, which are compatible with the LDO package. The DLMS is

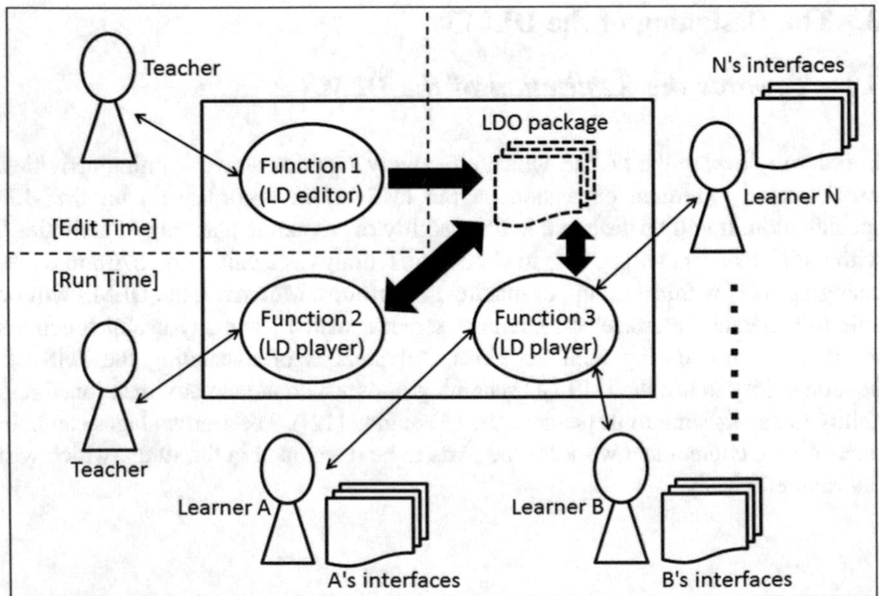

Fig. 3 DLMS's functions

applicable to two steps; Edit Time and Run Time as indicated in Fig. 3. In Edit Time, the teacher makes planning. (Then, the content of the LDO package will be created.) She/he can choose either a pair work or a group work, and later she/he can also change learners' roles in a group, or make new groups. Next, in Run Time, she/he chooses a pair work, for example, at a DM Point (Function 2), and assign who and who will become a pair. Then, "Interface A" will be generated on screens of a half of the learners. "Interface B" will be presented on those of the other half (Function 3). Note that the content of Interface A is different from that of Interface B because Learner A and Learner B have a different assignment in a pair work.

3.3 LDO_A Schema

LDO_A is created using IMS-LD Level A schema and recorded as Extensible Markup Language (XML) codes. The items to be created as LDO_A are defined as shown in Table 1. The [Learning Unit] consists of the [roles], [learning activities] and [support activities] and its structure is as described in Sect. 2.2. The base of a "flow of learning activities" is constructed as a successive structure of multiple [Learning Units]. The other elements other than those listed in Table 1 are used in the actual XML codes in the previous IMS-LD schema [1].

3.4 LDO_B Schema

To realize LDO_B, we attempt to expand the use of Level B [properties] and drawing up new schema that corresponds to the expansion, because LDO_B cannot be realized using the previous IMS-LD Level B schema.

Figure 4 is a new schema of LDO_B drawn in XML code with Document Type Definition (XML-DTD). It shows the schema used to describe information necessary to complete "the flow of learning activities combining multiple roles and groups" and to realize a "dynamic planning of activities" in XML codes. We name it "manage-role & group-rule (<manage-role-and-group-rule>)". The content described by the schema enables the DLMS to generate web interface for activities. Table 2 shows the elements and attributes used in the schema of LDO_B.

Table 1 LDO_A items

Description item	Corresponding IMS-LD element
[Learning Unit]	<act>
[roles: learner]	<learner>
[roles: staff]	<staff>
[learning activity]	<learning-activity>
[support activity]	<support-activity>

```
<!DOCTYPE manage-role-and-group-rule [
  <!ELEMENT manage-role-and-group-rule
     (target-act, decision-act, decision-role,
      role-property, group-property, title, explanation?,
      role-and-group-structures)>
  <!ATTLIST manage-role-and-group-rule
    identifier ID #REQUIRED>
  <!ELEMENT target-act EMPTY>
  <!ATTLIST target-act ref CDATA #REQUIRED>
  <!ELEMENT decision-act EMPTY>
  <!ATTLIST decision-act ref CDATA #REQUIRED>
  <!ELEMENT decision-role EMPTY>
  <!ATTLIST decision-role ref CDATA #REQUIRED>
   <!ELEMENT role-property EMPTY>
  <!ATTLIST role-property ref CDATA #REQUIRED>
  <!ELEMENT group-property EMPTY>
  <!ATTLIST group-property ref CDATA #REQUIRED>
  <!ELEMENT title (#PCDATA)>
  <!ELEMENT explanation (#PCDATA)>
  <!ELEMENT role-and-group-structures
     (role-and-group-structure+)>
  <!ELEMENT role-and-group-structure
     (title, structure-members,
      role-and-group-learning-objects?)>
  <!ATTLIST role-and-group-structure
    identifier ID #REQUIRED>
  <!ATTLIST role-and-group-structure
     appropriate-number-of-people NMTOKEN
     #REQUIRED>
  <!ELEMENT structur-members (role-of-member+)>
  <!ELEMENT role-of-member EMPTY>
  <!ATTLIST role-of-member ref CDATA #REQUIRED>
  <!ATTLIST role-of-member
     appropriate-number-of-people NMTOKEN
     #REQUIRED>
  <!ELEMENT role-and-group-learning-objects
     (role-and-group-learning-object+)>
  <!ELEMENT role-and-group-learning-object (title, item)>
  <!ATTLIST role-and-group-learning-object
    identifier ID #REQUIRED>
  <!ATTLIST role-and-group-learning-object
     type CDATA #IMPLIED>
  <!ELEMENT item EMPTY>
  <!ATTLIST item identifier ID #REQUIRED>
  <!ATTLIST item identifierref CDATA #REQUIRED>
  <!ATTLIST item parameters CDATA #IMPLIED>
]>
```

+ : Repetition of 1 or more , ? : 0 or 1

Fig. 4 Schema of LDO_B

Table 2 Details of <manage-role-and-group-rule> in LDO_B

Elements/attributes		Details of elements/attributes
<manage-role-and-group-rule>		A collection of information that is necessary to control learning activities, roles and groups
attributes	Identifier	Describes ID (identifier) of the <manage-role-and-group-rule>
<target-act>		Specifies ID of the "Learning Unit" to be controlled
attributes	Ref	
<decision-act>		Specifies ID of the "Learning Unit" before the "DM Point" is provided
attributes	ref	
<decision-role>		Specifies ID of the "role" which controls learning activities, roles and groups at the "DM Point"
attributes	ref	
<role-property>		Specifies ID of the property to store each learner's "role" information
attributes	ref	
<group-property>		Specifies ID of the property to store each learner's "group" information
attributes	ref	
<title>		Describes the title
<explanation>		Describes the explanation (can be omitted)
<role-and-group-structures>		A collection of <role-and-group-structure>
<role-and-group-structure>		Defines the "role and group structure"
attributes	identifier	Describes the "role and group structure" ID
	appropriate-number-of-people	Specifies the optimal number of users for the "role and group structure"
<structure-members>		Defines the "roles" that constitute the "role and group structure"
<role-of-member>		Specifies ID of the "roles" that constitute the "role and group structure"
attributes	ref	
	appropriate-number-of-people	Specifies the optimal number of users for each "role" that constitutes the "role and group structure"
<role-and-group-learning-objects>		A collection of <role-and-group-learning-object> (can be omitted)
<role-and-group-learning-object>		Defines digital contents to be used for each "role and group structure"
attributes	identifier	Describes the <role-and-group-learning-object> ID
	type	Describes the type of <role-and-group-learning-object> (can be omitted)
<item>		Defines resources used at <role-and-group-learning-object>
attributes	identifier	Describes the <item> ID
	identifierref	Specifies ID of the resource to be used
	parameters	Sets the parameters to be handed over to the resources (can be omitted)

```
<manage-role-and-group-rule identifier="rule-id-1">
  <target-act ref="Learning Unit 3"/>
  <decision-act ref="Learning Unit 2"/>
  <decision-role ref="Teacher"/>
  <role-property ref="Property 1"/>
  <group-property ref="Property 2"/>
  <title>"Individual Tasks" or "Group Tasks"</title>
  <role-and-group-structures>
   <role-and-group-structure identifier="rgs-id-2" appropriate-number-of-
       people="1">
    <title>Individual Tasks</title>
     <structure-members>
      <role-of-member ref="Learner (individual)"
          appropriate-number-of-people="1"/>
     </structure-members>
     <role-and-group-learning-objects>
      <role-and-group-learning-object identifier="rglo-id-2">
       <title>Materials for Individual Learning</title>
       <item identifier="item-id-2" identifierref="r-id-2">
      </role-and-group-learning-object>
     </role-and-group-learning-objects>
   </role-and-group-structure>
   <role-and-group-structure identifier="rgs-id-3" appropriate-number-of-
       people="3">
    <title>Group Tasks</title>
     <structure-members>
      <role-of-member ref="Staff 1"
          appropriate-number-of-people="1"/>
      <role-of-member ref="Staff 2"
          appropriate-number-of-people="1"/>
      <role-of-member ref="Leader "
          appropriate-number-of-people="1"/>
     </structure-members>
     <role-and-group-learning-objects>
      <role-and-group-learning-object identifier="rglo-id-3">
       <title>Materials for Group Learning</title>
       <item identifier="item-id-3" identifierref="r-id-3">
      </role-and-group-learning-object>
     </role-and-group-learning-objects>
   </role-and-group-structure>
   <role-and-group-structure identifier="rgs-id-4" appropriate-number-of-
       people="1">
    <title>Reflecting</title>
     <structure-members>
      <role-of-member ref="Learner (reflection)"
          appropriate-number-of-people="1"/>
     </structure-members>
   </role-and-group-structure>
  </role-and-group-structures>
</manage-role-and-group-rule>
```

Fig. 5 Example of LDO_B

3.5 *Example of How to Describe LDO_B*

Figure 5 is an example of description of <manage-role-and-group-rule> of LDO_B. It shows a description which enables the teacher to decide the [roles] and [groups] of each learner ([learning activities management]) at the "DM Point" between Learning Units 2 and 3, as shown in Fig. 1.

As described in Sect. 2.3, the [learning activities] are associated with the [roles]. Therefore, <target-act> in Fig. 5 describes "Learning Unit 3", <decision-act> describes "Learning Unit 2" and <decision-role> describes "Teacher". Figure 5 describes two <role-and-group-structure> in order to deal with the split between "Individual Tasks" and "Group Tasks" in Fig. 1. It also has one <role-and-group-structure> in order to carry out "Learner (reflection)" by multi-roles. The LMS provides the interface that reads these descriptions and enables the teacher to assign each learner to "Individual Tasks" or "Group Tasks", and to specify multi-roles. The teacher determines each learner' s [roles] and [groups] on this interface. Everyone will be attributed with information about the [roles] and [groups] as determined by the teacher and the information is stored in "Property 1" as described in <role-property> and in "Property 2" as described in <group-property> (each property is instantiated by the learner). Furthermore, in addition to the resources already registered in [Learning Units], <role-and-group-learning-object> described in <role-and-group-structure> provides the learners with resources that are instantiated by the group, or by individual in case of self-learning.

The <manage-role-and-group-rule> schema solves the problems with IMS-LD pointed out in Sect. 2.3. A teacher's "dynamic planning of activities" can be realized by describing <manage-role-and-group-rule> in the DLMS.

4 The Development of the DLMS

4.1 *System Configuration*

The DLMS was developed in the free software package of XAMPP environment on Windows 8.1 Pro. It was developed as a web application, which works on Apache HTTP Server, and PHP was used as the development language.

The DLMS is equipped with the "LD editor" and "LD player" that are compatible with the LDO package as shown in Fig. 6. Information to be required to operate the "LD editor" and "LD player" is stored in the database (MySQL). In addition, in parts of analytical process of XML codes in the "LDO package", the "LD engine" (Copper Core [12]) is used. These linked operations support a dynamic planning of activities. The next section provides an example of how the DLMS works.

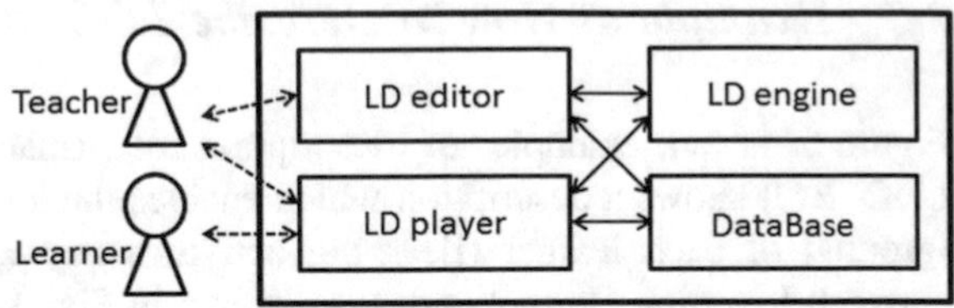

Fig. 6 System configuration of the DMLS

4.2 Example of System Operation

The teacher creates an LDO package using the LD editor (Fig. 7). (The example of the LDO package containing XML codes is shown in Fig. 5). Then she/he determines each learner's group and role on the LD player (Fig. 8). Figures 9 and 10 show the LD player interface of the learners view. The learner's interface is generated, based on the teacher's assignment of groups and roles. Here in the Learning Unit 3, each learner has two roles. In a group work, each learner has a role of either "Leader", "Staff 1" or "Staff 2". In addition to that everyone has a role of "Learner (reflection)".

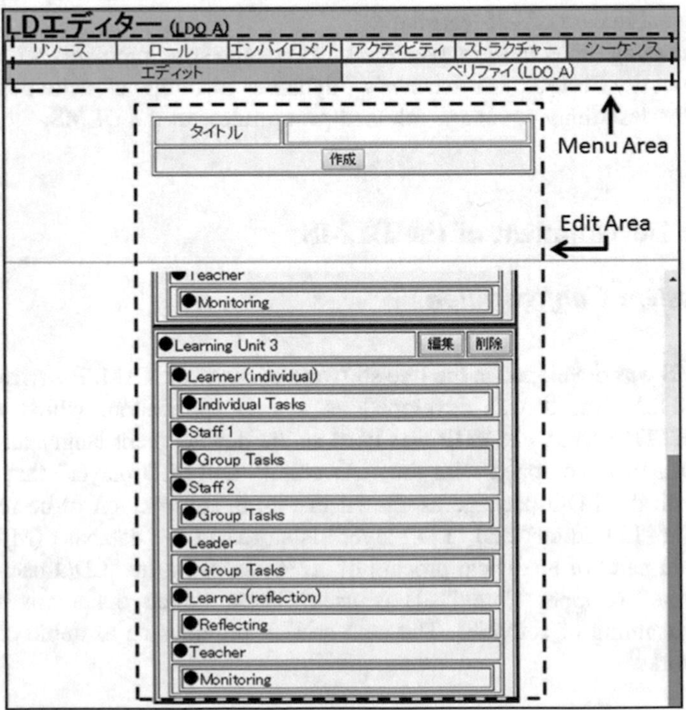

Fig. 7 Screenshot of LD editor

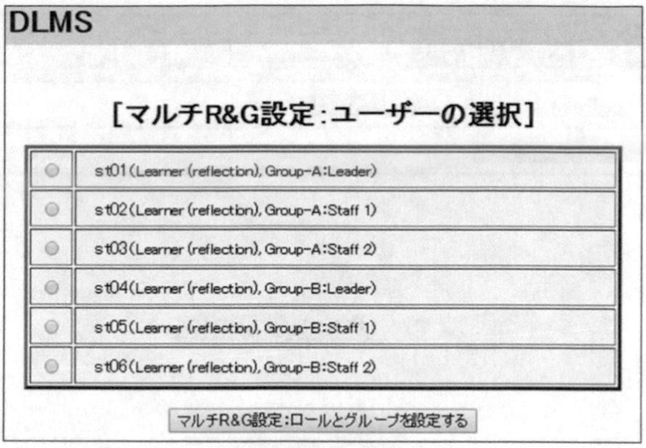

Fig. 8 Learner's group & role management screen

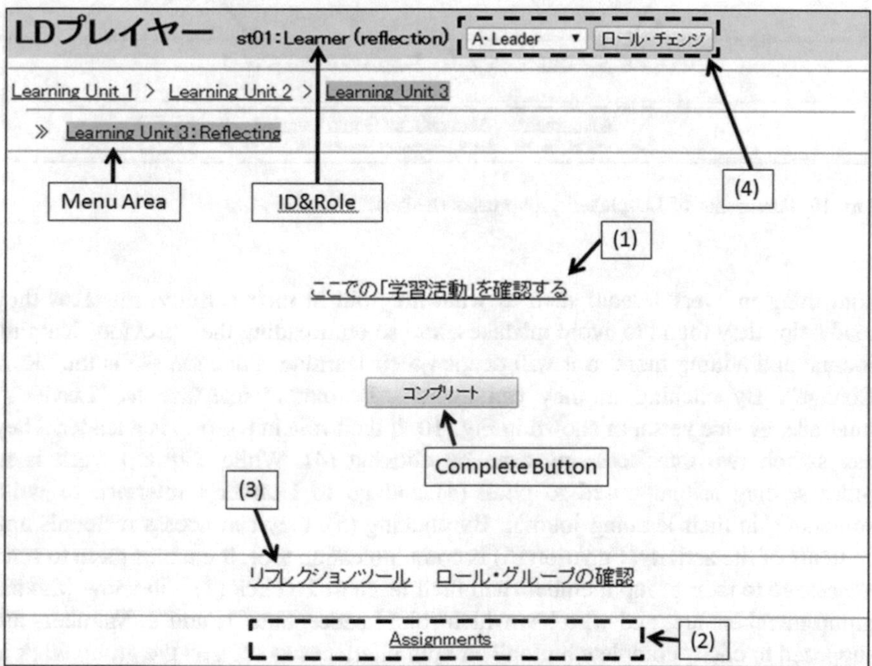

Fig. 9 Screenshot of LD player: learner (reflection)

Figure 9 is the "Learner (reflection)" interface. By clicking function (1), learners can look at the content of activity. By clicking function (2), they can access links to necessary contents. Function (3) is the reflection tool. It has a function of attaching files, in which learners keep the learning journal. They are supposed to write down

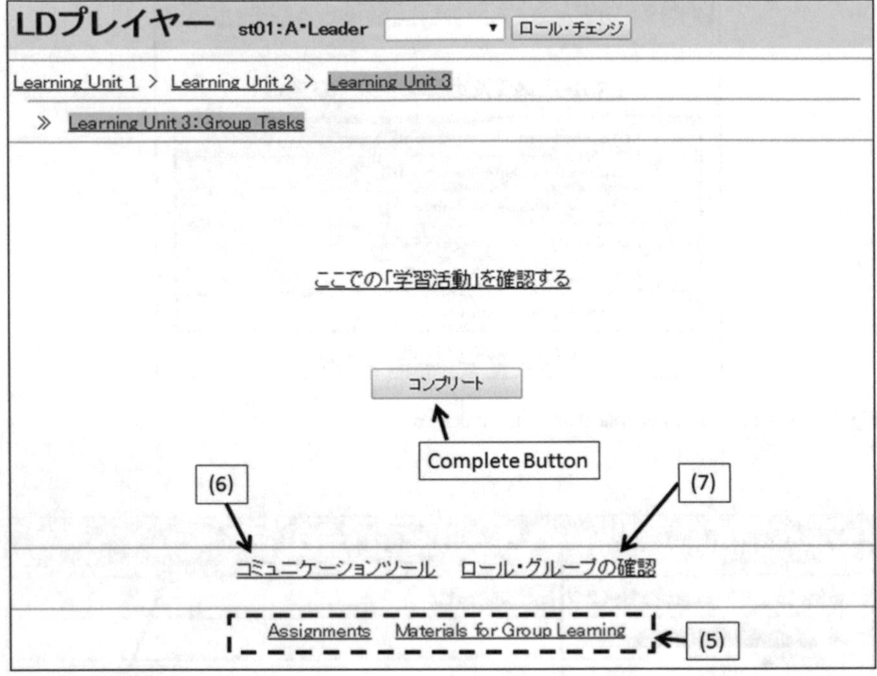

Fig. 10 Screenshot of LD player: group tasks (leader)

something in every lesson, such as what they did in their activity, mistakes they made, tips they found to avoid mistakes, and so on. Reading their previous learning journal and adding more to it will deepen their learning. Function (4) is the "Role Change". By clicking it, they can switch "Learner's" interface to "Leader's" interface, or vice versa, as shown in Fig. 10, if their role in a group is a leader. They can switch two interfaces any time by clicking (4). While a group work is in process, they are supposed to click (4) and go to Learner's interface to write something in their learning journal. By clicking (5), they can access materials and contents of the activity. Function (6) is communication tool. It enables them to send a message to their group members and their teacher. To click (7) will show who his group members are, and who has which role; Leader, Staff 1, and 2. Members are supposed to click "complete button" on both interfaces as soon as the group work is finished. If their teacher has planned a role-change within a group next, interface of "Staff 1" will be generated on the screen of a learner who had "Leader's" interface. In this way, the DLMS supports "dynamic planning of activities".

5 Conclusion

In this study, we have proposed the "LDO package" to be the basis of the DLMS, and explained how to develop the DLMS which supports "dynamic planning of activities". We also discussed the DLMS with containing LD editor and LD player, which are compatible with the "LDO package". Future tasks will be to implement the DLMS in classes, and to work on system evaluation.

References

1. IMS Global Learning Consortium. Learning Design Specification (2003). IMS Global Learning Consortium. http://www.imsglobal.org/learningdesign/index.html. Accessed 10 Apr 2016
2. Fink, L.D.: Creating Significant Learning Experiences: An Integrated Approach to Designing College Courses. Wiley, San Francisco (2013)
3. Guerrero-Roldán, A.E., García-Torà, I., Prieto-Blázquez, J., Minguillón, J.: Using an IMS-LD based questionnaire to create adaptive learning paths. In: Frontiers in Education Conference (FIE), pp. F1 J-1–F1 J-6. IEEE (2010)
4. Mavroudi, A., Hadzilacos, T.: Implementation of adaptive learning designs. Ann. Univ. Craiova, Ser. Autom. Comput. Electron. Mech. **9**(2), 18–24 (2012)
5. Specht, M., Burgos, D.: Modeling adaptive educational methods with IMS learning design. J. Interact. Media Educ. **2007**(1) (2007). doi:10.5334/2007-8
6. Towle, B., Halm, M.: Designing adaptive learning environments with learning design. In: Koper, R., Tattersall, C. (eds.) Learning Design. Springer, Berlin (2005)
7. Gorissen, P., Tattersall, C.: A learning design worked example. In: Koper, R., Tattersall, C. (eds.) Learning Design. Springer, Berlin (2005)
8. Hernández-Leo, D., Villasclaras-Fernández, E.D., Asensio-Pérez, J.I., Dimitriadis, Y., Jorrín-Abellán, I.M., Ruiz-Requies, I., Rubia-Avi, B.: COLLAGE: a collaborative learning design editor based on patterns. Educ. Technol. Soc. **9**(1), 58–71 (2006)
9. Koper, R., Miao, Y.: Using the IMS LD standard to describe learning designs. In: Lockyer, L., Bennett, S., Agostinho, S., Harper, B. (eds.) Handbook of Research on Learning Design and Learning Objects: Issues, Applications and Technologies. Information Science Reference, Hershey (2008)
10. Villasclaras-Fernández, E.D., Hernández-Gonzalo, J.A., Leo, D.H., Asensio-Pérez, J.I., Dimitriadis, Y.A., Martínez-Monés, A.: InstanceCollage: a tool for the particularization of collaborative IMS-LD scripts. Educ. Technol. Soc. **12**(4), 56–70 (2009)
11. LAMS Foundation. LAMS 2.4. LAMS Foundation (2012). http://lamsfoundation.org/. Accessed 10 Apr 2016
12. CopperCore Project. CopperCore 3.3. SourceForge (2008). http://coppercore.sourceforge.net/index.shtml. Accessed 10 Apr 2016

Content-Based Microscopic Image Retrieval of Environmental Microorganisms Using Multiple Colour Channels Fusion

Yanling Zou, Chen Li, Kimiaki Shiriham, Florian Schmidt, Tao Jiang and Marcin Grzegorzek

Abstract *Environmental Microorganisms* (EMs) are usually unicellular and cannot be seen with the naked eye. Though they are very small, they impact the entire biosphere by their omnipresence. Traditional DeoxyriboNucleic Acid (DNA) and manual investigation in EMs search are very expensive and time-consuming, we develop an EM search system based on Content-based Image Retrieval (CBIR) method by using multiple colour channels fusion. The system searches over a database to find EM images that are relevant to the query EM image. Through the CBIR method, the features are automatically extracted from EM images. We compute the similarity between a query image and EM database images in terms of each colour channel. As many colour channels exist, a weight fusion of similarity in different channels is required. We apply Particle Swarm Optimisation (PSO), Fish Swarm Optimisation Algorithm (FSOA), Invasive Weed Optimization (IWO) and Immunity Algorithm (IA) to laugh fusion. Then obtain the re-weighted EM similarity and final retrieval result. Experiments on our EM dataset show the advantage of the proposed multiple colour channels fusion method over each single channel result.

Keywords Environmental microorganism · Multiple colour channels · Optimisation-based fusion

1 Introduction

Environmental Microorganisms (EMs) are present in every part of the biosphere (rivers, forests, mountains, etc.), playing critical roles in earth's biogeochemical cycles [1]. They are cost-effective agents for in-situ remediation of domestic, agricultural and industrial wastes and subsurface pollution in soils, sediments and marine

Y. Zou (✉) · T. Jiang
Chengdu University of Information Technology, Chengdu, China
e-mail: zyl@cuit.edu.cn

C. Li · K. Shiriham · F. Schmidt · M. Grzegorzek (✉)
Institute for Vision and Graphics, University of Siegen, Siegen, Germany
e-mail: marcin.grzegorzek@uni-siegen.de

R. Lee (ed.), *Computer and Information Science*,
Studies in Computational Intelligence 656, DOI 10.1007/978-3-319-40171-3_9

environments. For example, Vorticella, which can digest organic pollutant in wastewater and improve the quality of fresh water. Since traditional DeoxyriboNucleic Acid (DNA) and manual investigation in EM search are very expensive and time-consuming. To enhance the effectiveness of EM information search, we are developing EM search systems based on Content-Based Image Retrieval (for short, EM-CBIR) [2, 3]. The system searches over a database to find EM images that are relevant to the query EM image. Through the CBIR method, the features are automatically extracted from EM images to represent their semantic properties, such as colour, shape and texture, etc.

Since the colour information of EM images depend on different light sources of various microscopes, it is not sufficient to use only one single colour channel feature for EM search. Hence, we decompose each EM image to multiple colour channels, including 10 as follows: RGB, R, G, B, Grey-level (I), HSV, H, S, V and mean HSV

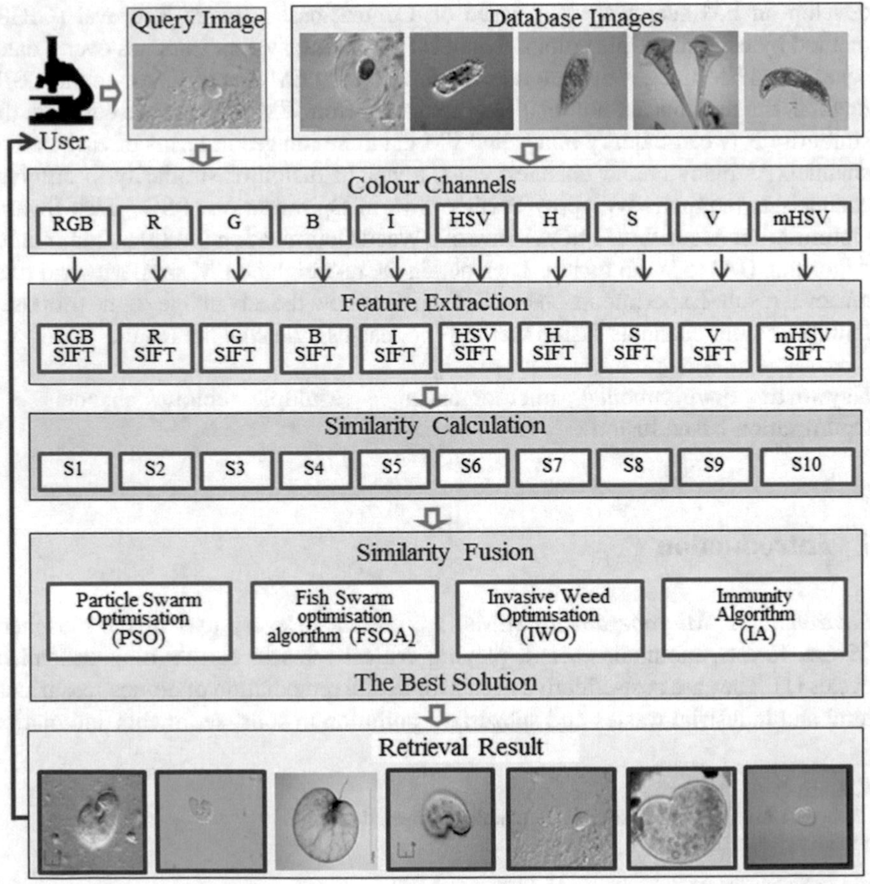

Fig. 1 Work flow of the EM-CBIR system

(mHSV). Then, we extract *Scale-invariant Feature Transform* (SIFT) [4] features individually from each of these colour channels, because they can capture detailed appearances of an EM and is useful for achieving accurate matching. Moreover, in order to enhance the performance of EM retrieval, we use *Fish swarm optimisation algorithm* (FSOA) [5], *Particle Swarm Optimisation* (PSO) [6, 7], *Invasive Weed Optimisation* (IWO) [8] and *Immune Algorithm* (IA) [9] to obtain the re-weighted EM similarities. Built on these different weighted similarities, we adopt a late fusion approach to combine retrieval results, and apply *Average Precision* (AP) and mean AP (mAP) to evaluate the retrieval result.

Figure 1 illustrates the pipeline of the proposed EM-CBIR system. At first, a user give a query image, the system conducts image initialisation to decompose it into 10 different colour channels. Then, features are extracted from each channel. Furthermore, we employ a similarity measure to generate matching features and detect how many features are matched between the query image and EM database images. Afterwards, by comparing 4 optimisation methods, we obtain the best re-weighted EM similarity. Finally, the retrieval system outputs similar images as result feedback to the query sorted from most similar to dissimilar.

2 Related Work

Nowadays there are various methods for image retrieval and classification of microorganism. The following section will discuss the most relevant. To identify the rich marine life resources the marine microorganism are important. Sheikh et al. [10] developed a system for image retrieval and evaluated the effectiveness of different kinds of image descriptors for indexing marine life images, where features are represented in shape, color and texture. Another kind of microorganism are medical microorganism. They are used for the prevention, diagnosis and treatment of infectious diseases. To archive and retrieve histopathology images by content, Caicedo et al. [11] proposed a framework to automatically annotate images and can recognize high-level concepts after analyzing visual image contents.

Akakin and Gurcar [12] developed a content-based microscopic image retrieval system that utilises a reference database containing images of more than one disease. In this research, a multi-tiered method is used to classify and retrieve microscopic images involving their specific subtypes, which are mostly difficult to discriminate and classify. The system enables both multi-image query and slide-level image retrieval. Previous work for EM are investigated in [2, 3, 13–15]. A semi-automatic image segmentation approach is proposed to refine an EM region. Then, this region is represented by several shape descriptors. Support vector machines are used to discriminate between regions of an interesting EM and the others.

All of the aforementioned methods use image segmentation and shape descriptors to extract features of microorganism images. Although their effectiveness depends on the accuracy of image segmentation, it is still difficult to achieve accurate segmentation. Li et al. [16] introduces an approach for EM classification using Sparse

Coding (SC), which extracts sufficient local features from an image and reconstructs it by a sparse linear combination of bases, and Weakly Supervised Learning (WSL) which jointly performs the localisation and classification of EMs by examining the local information in training images. Since shape features need segmentation and SC is computationally expensive, we particularly choose to use SIFT features [4]. SIFT features are usually extracted from grey images; this ignores the useful colour information which finally hinders the ultimate performance. We decompose an EM image into different colour channels. Bosch [17] extracted SIFT features over three channels of HSV colour space. van de Sande et al. [18] extended several methods to effectively make use of colour spaces. While [12, 13] use various descriptors, for EMs, our method extracts SIFT features from different colour spaces is required. We experimentally show the effectiveness of using multiple colour spaces.

In our work, a late fusion scheme is implemented. It exploits the similarity vectors from different colour channels separately and combines the retrieval results. There is another common scheme, early fusion, which combines all similarities into a single high-dimensional vector. However, some important vectors may be weakened when it performs retrieval tasks based on this single vector. Thus, we have decided to apply late fusion in our approach. There are several optimisation algorithms inspired in the last years, such as FSOA, PSO, IWO and IA. In order to choose the best solution, we investigate these four methods to compute weights for similarities in terms of different colour channels.

The FSOA was based on the behavior of fish swarm in search for food [5]. From the optimisation point of view, this behavior is associated to learning capacity that can lead the fish swarm to new directions. For example, the exploration of new food sources (design space). Each fish represents a candidate solution of the optimisation problem. In an optimisation problem, the amount of food in a region is inversely proportional to value of objective function. The aquarium is the design space where the fish can be found. It can achieve faster convergence speed and require few parameters to be adjusted. The PSO was originally developed by Kennedy and Eberhart in 1995 [6] and is an evolutionary computation algorithm that can be applied to solve complex optimisation problems [7]. It is conducted as a stylised representation of the movement of bird flock, where each particle keeps track of its coordinates in the solution space associating with the best solution. During the iterations, the particles are updated by the particle's personal best value (a relevant position) and the swarm's global best value (the best position among all the solutions). These positions represent the solution of the problem. PSO has been applied in image retrieval as well [19]. It is characterised by a low computational complexity and fast convergence, and provides a way to solve many highly non-linear and multi-dimensional problems. The IWO inspired from the phenomenon of colonization of invasive weeds in nature, is based on weed biology and ecology. It has been shown that capturing the properties of the invasive weeds, leads to a powerful optimisation algorithm [8]. The IA is theoretically to utilize the locally characteristic information for seeking the ways and means of finding the optimal solution when dealing with difficult problems. To be exact, it utilizes the local information to intervene in the globally parallel process and restrain or avoid repetitive and useless work during the course, so as to overcome the blindness in action of the crossover and mutation [20].

3 EM-CBIR Methods

In this section, the proposed EM-CBIR method is presented in detail. First, each EM image is decomposed into images on different colour channels. Then, we extract SIFT features and compute the similarity between sets of SIFT descriptors. Finally, applying four optimisation methods to re-weight the similarities.

3.1 *Multiple Colour Channels and Feature Extraction*

Most EM images show different aspects depending on the image between light sources of varied microscopes. Thus, a colour channel plays a vital role for the performance of EM image search. It can be processed using RGB colour space or HSV colour space. In situations where colour description plays an integral role, HSV is often preferred over RGB. RGB defines a colour in terms of a combination of primary colours (i.e., Red, Green and Blue), whereas HSV describes a colour using more intuitive comparisons such as hue, saturation and brightness. In this case, we decompose EM images into 10 colour channels to increase discriminative power [21, 22]. For example, there are 10 different colour channels shown in Fig. 2. Then, we extract SIFT features on each channel to build an EM image feature matrix. The extraction of SIFT features does not need any pre-processing such as segmentation. In Fig. 2, a yellow dot is shown at the position of each interest point. In our method, the SIFT feature which results in a 128-dimensional vector, is separated to represent the statistical information of each channel.

The working process of feature extraction can be briefly sketched as follows. Firstly, we use a Differences-of-Gaussian (DoG) [23, 24] function to identify poten-

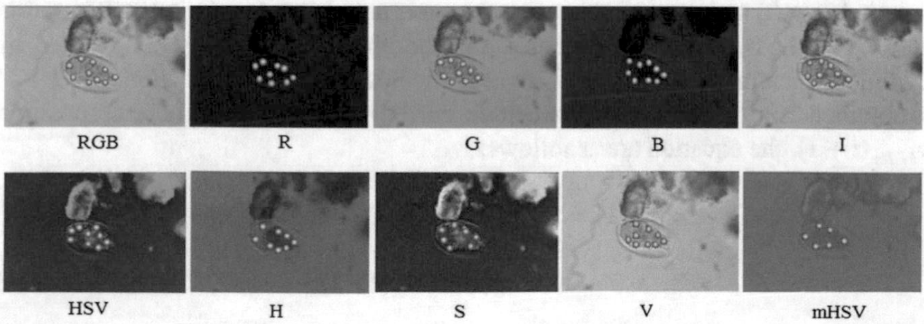

Fig. 2 Different colour channels with interest points of EM image

tial interest points on each colour channel of EM image. Secondly, interest points are selected based on measures of their stability. Thirdly, the algorithm assigns orientations to the interest points. At last, the EM image gradients are measured at the selected scale in the region around each interest point. These are transformed into a representation that allows for significant levels of local shape distortion and change in illumination.

We match SIFT features extracted from two images using Harris Corner Detector [25] and RANSAC [4]. The overall similarity between these images is computed by summing Euclidean distances between matched SIFT features. To avoid any deviations from different value ranges of different features, we normalise each feature, so that the values range from 0 to 1. Since using a single channel features is obviously insufficient for achieving the state-of-the-art performance, we fuse similarities calculated on different features. Here, we define similarity as $s = 1 - D$, where $D \in [0, 1]$ is the normalised Euclidean distance between two images. Suppose N is the number of channels. We initialise the similarity groups as $s_1, s_2, \dots, s_N$. The fused similarity S is calculated by:

$$S(N) = \frac{1}{N} \sum_{i=1}^{N} x_i s_i \quad , \tag{1}$$

where x_i is the weight which indicates the usefulness of the corresponding similarity s_i.

3.2 Late Fusion and Retrieval Approach

In order to train the similarity matrix and obtain the weights, we first use four optimisation methods. Then we applied a late fusion approach to combine retrieval results. The weight computation is formulated as an optimisation process.

Fish Swarm Optimisation Algorithm **(FSOA)**

We suppose each weight (a swarm position) is represented as x_n. The current position is noted as $x_n(t)$ and the next selected position is noted as $x_n(t+1)$. At each iteration step $t+1$, the behaviors are executed and the *i*th position $x_n(t+1)$ is updated by adding an increment $\Delta x_n(t+1)$ which only includes optimal information. $\Delta x_n(t+1)$ is obtained from the difference between the current position $x_n(t)$ and a better position $x_{better}(t+1)$, the equation are as follows:

$$\Delta x_n(t+1) = r_1 f_{step}(x_{better}(t+1) - x_n(t)) \quad , \tag{2}$$

$$x_n(t+1) = x_n(t) + \Delta x_n(t+1) \quad , \tag{3}$$

where the r_1 is an uniform random sequences in $[0, 1]$, f_{step} is the moving step length.

***Particle Swarm Optimisation* (PSO)**

At first, we have to define the initial parameters of the particles. Suppose each weight (the swarm particle) is represented as x_n. The best personal value (a relevant position) of the nth weight is represented as p_n, and the global best (the best position) is noted as g^t. Then, we define the next position of each weight by adding a speed to the current position. Here, the speed of each weight is set as follows:

$$v_n^t = \psi v_n^{t-1} + c_1 r_1 (p_n^t - x_n^{t-1}) + c_2 r_2 (g^t - x_n^{t-1}) \quad , \tag{4}$$

where t is the current iteration number; c_1 and c_2 are learning factors in the range $[0, 2]$; r_1 and r_2 are uniform random sequences in $[0, 1]$; ψ is an inertial weight parameter, progressively decreasing along iterations [26]. In this case, ψ decreases proportionally to the number of corresponding weights. Finally, the weight is updated according to the following formulas:

$$x_n^t = x_n^{t-1} + v_n^t \quad . \tag{5}$$

***Invasive Weed Optimisation* (IWO)**

Since an array antenna with elements, separated by a uniform distance, the normalized weight factor is given by:

$$AF(\theta) = \frac{1}{AF_{max}} \sum_{n=1}^{n_1} I_n e^{j2\pi n d sin\theta / \lambda} \quad , \tag{6}$$

where AF_{max} is the maximum value of the magnitude of the weight factor, I_n is amplitude coefficient, θ is the angle from the normal to the array axis, d is assumed to be $\lambda/2$, λ is the wavelength.

***Immune Algorithm* (IA)**

We suppose each weight (a vaccination) is represented as x_n. Selecting a x_n in the present weight $E_l = (x_1, x_2, \dots, x_{n_0})$ to join in the new parents with the probability below:

$$P(x_n) = \frac{e^{f(x_n)/T_l}}{\sum_{n=1}^{n_0} e^{f(x_n)/T_l}} \quad , \tag{7}$$

where $f(x_n)$ is the fitness of the individual x_n, $\{T_l\}$ is the temperature-controlled series approaching 0.

3.3 Evaluation and Termination Criteria

Based on the computation of similarities, we select AP to measure the retrieval performance. AP has been developed in the field of information retrieval, and is used to evaluate a ranked list of retrieved samples [27]. In our EM-CBIR system, AP is defined as follows:

$$\mathrm{AP} = \frac{\sum_{k=1}^{m}(P(k) \times \mathrm{rel}(k))}{M}, \tag{8}$$

where M is the number of relevant EM images, $P(k)$ is the precision by regarding a cut-off position by the kth position in the list, and $\mathrm{rel}(k)$ is an indicator function which takes 1 if the EM image ranked at the kth position is relevant, otherwise 0. Thus, AP represents the average of precisions, each of which is computed at the position where a relevant EM image is ranked. The value of AP increases when relevant EM images are ranked at higher positions. In addition, as our experiments are performed over 21 classes of EM images, we apply mean AP (mAP) to aggregate the AP of the individual classes. It is calculated by taking the mean value of APs.

The weights matching process terminates when a predefined number of iterations is reached. Afterwards, we obtain the re-weighted EM similarity and all relevant EM images are shown from similar to dissimilar to the user.

4 Experimental Results

We conduct experiments on a real EM dataset (EMDS) containing 21 classes of EMs $\{\omega_1, \dots, \omega_{21}\}$ as shown in Fig. 3. Each class is represented by 20 microscopic images. In our experiments, we use each EM image as a query image once and all the remaining images for testing. As RGB and HSV are the two most common representations in the colour model, the EM-CBIR system extracts SIFT features from different colour channels of the EM image which include 10-dimensional colour channels (RGB, R, G, B, I, HSV, H, S, V and mean HSV). Then, we calculate the similarity and obtain the similarity matrix. Further, four optimisation approaches are used to re-weight similarities. Finally, the result of the best solution are applied for launching late fusion. The performance of our system was evaluated in terms of AP and mAP for all 21 classes.

In order to evaluate the effectiveness of multiple colour channels, we compare independent colour channels information of EM images for further experiments. The chart in Fig. 4 compares the average percentage precision of retrieval result for PSO, FSOA, IWO and IA. The result of mAP using PSO is 23 %. The FSOA yields the mAP of 18.16 %, which obtains an improvement for ω_{15}, ω_{16} and ω_{19}. The IWO achieves a mAP of 15.04 %, ω_{12} is improved. It is much lower than PSO, FSOA and IA. The mAP of IA is 19.86 %, which obtains an improvement for ω_{16} and ω_{17}. Thus, the performance of PSO obviously outperforms the other three methods. For detail, Fig. 5 compares the AP of the 10 channel categories and fusion using PSO on 21 classes of EM images, as well as mAP. Table 1 shows the comparison between PSO-based fusion and the best single results. It is observed that PSO-based fusion method is much higher than the best single channel features of 17 %. We discover that our

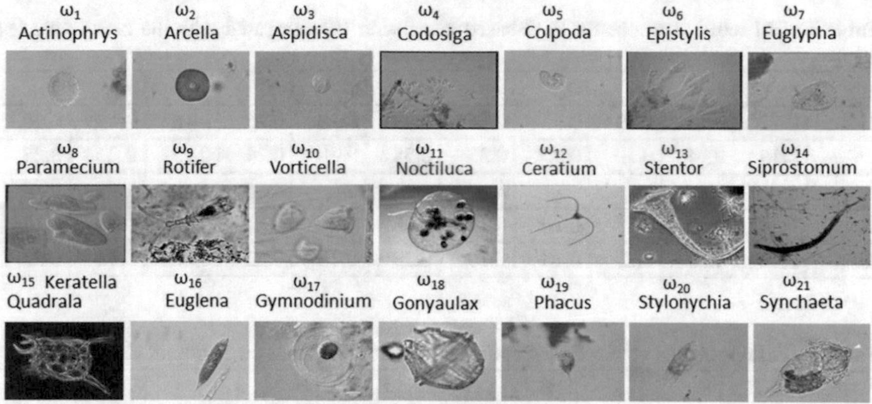

Fig. 3 EM categories of EMDS

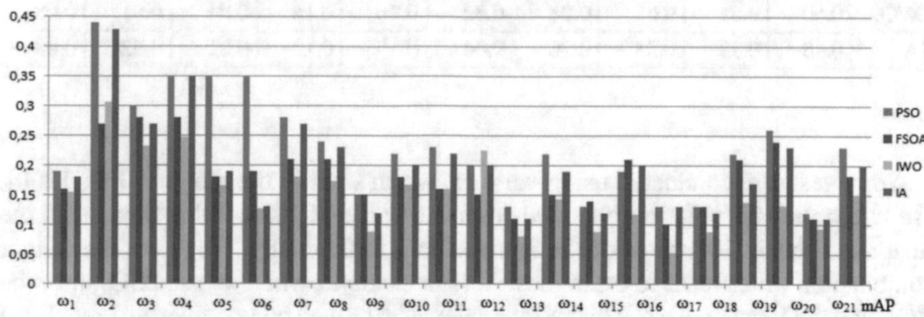

Fig. 4 Comparison between the performances of the PSO, FSOA, IWO and the IA

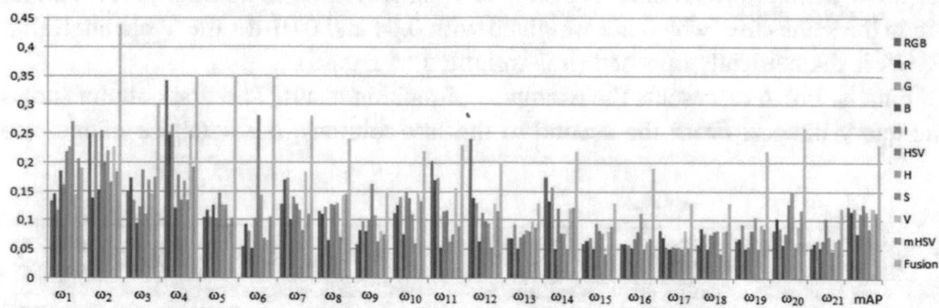

Fig. 5 Evaluation of 10 colour channels and fusion using PSO on 21 classes of EM images

method improves the retrieval results for 18 classes. In particular, it yields an outstanding performance improvement for $\omega_2, \omega_4, \omega_5$ and ω_6, which has fewer interest points only with a single colour channel. However, there is no further enhancement for ω_{13} and ω_{21}. By analysing the EM images, we find that the performance is worse for the EM which has high transparency, rough edge or more impurities, such as ω_{13} and ω_{17} in Fig. 3.

Table 1 EM retrieval results for PSO-based late fusion (F) compared with the best single (S) results

	ω_1	ω_2	ω_3	ω_4	ω_5	ω_6	ω_7	ω_8	ω_9	ω_{10}	ω_{11}
S	0.27	0.26	0.19	0.35	0.15	0.28	0.18	0.15	0.16	0.15	0.20
F	0.19	0.44	0.30	0.35	0.35	0.35	0.28	0.24	0.15	0.22	0.23
	ω_{12}	ω_{13}	ω_{14}	ω_{15}	ω_{16}	ω_{17}	ω_{18}	ω_{19}	ω_{20}	ω_{21}	mAP
S	0.24	0.10	0.18	0.10	0.11	0.08	0.09	0.11	0.15	0.10	0.17
F	0.20	0.13	0.22	0.13	0.19	0.13	0.13	0.22	0.26	0.12	0.23

Table 2 Fusion weights obtained by 4 optimisation-based methods on 10 colour channels

	RGB	R	G	B	I	HSV	H	S	V	mHSV
PSO	0.72	0.08	0.46	0.34	0.43	0.25	0.16	0.06	0.86	0.09
FSOA	0.04	0.36	0.08	0.57	0.13	0.35	0.48	0.88	0.59	0.50
IWO	0.60	0.61	0.68	0.28	0.82	0.26	0.16	0.04	0.81	0.08
IA	0.45	0.27	0.33	0.30	0.46	0.19	0.12	0.01	0.92	0.08

Some tests are conducted using various swarm sizes in the range of [50, 1200]. We chose the best solution obtained from the value of 1000. Table 2 presents the fusion weights in 10 groups of the four methods individually. We select the weight combination which leads to the highest AP. For example, when we use PSO, since the RGB and the V channel work better for retrieval tasks than others, they are associated with the weights 0.86 and 0.72, respectively. It is also noted that PSO works not good for the S channel, which obtains a smaller weight of 0.06. In addition, IWO and IA are in the same case, which are weighted with 0.04 and 0.01 for the V channel. But, FSOA is diametrically opposed (it is weighted with 0.88).

Finally, Fig. 6 represents the examples of ranking result. The first column shows the query images. From the second to the last columns, the database images are

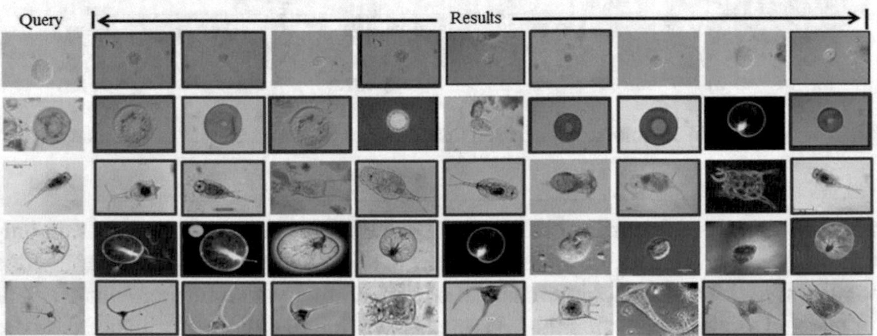

Fig. 6 Examples of EM retrieval results using multiple colour channels fusion

sorted by their SIFT similarities from similar to dissimilar. The images in red boxes are the relevant images. The experiment validates its usefulness for EM retrieval. This convincing result indicates that our late fusion method is necessary to improve the overall performance.

5 Conclusion and Future Work

In this paper, we introduced an EM-CBIR system using multiple colour channels fusion. In order to enhance the final retrieval performance by generating semantically meaningful representation, we decompose each EM image into different colour channels. Then, we extract features from each channel, and use Euclidean distance to measure the similarity of EM images. Afterwards, we propose a late fusion method using PSO, FSOA, IWA and IA to obtain the best computation weights. This enables us to calculate the re-weighted EM similarity in the retrieval task. Experimental results using optimisation-based fusion on multiple colour channels are much better than those of using single channel retrieval evaluation, especially PSO. This validates the effectiveness of the proposed methods. In our future work, we will address RGB-SIFT to improve the feature extracting performance of EM images [18].

Acknowledgments Research activities leading to this work have been supported by Program of Study Abroad for Young Scholar of CUIT, China Scholarship Council (CSC), project (No. 2015GZ0197, 2015GZ0304) Supported by Scientific Research Fund of Sichuan Provincial Science & Technology Department. We also thank Prof. Zhongren Guan, Prof. Rui Fang and B.A. Kristin Klaas for their technological help.

References

1. Pepper, I.L., Gerba, C.P., Gentry, T.J.: Environmental Microbiology. Academic Press, San Diego, USA (2014)
2. Zou, Y., Li, C., Boukhers, Z., Shirahama, K., Jiang, T., Grzegorzek, M.: Environmental microbiological content-based image retrieval system using internal structure histogram. In: Proceedings of CORES (2015)
3. Li, C., Shirahama, K., Grzegorzek, M.: Application of content-based image analysis to environmental microorganism classification. Biocybern. Biomed. Eng. **35**(1), 10–21 (2015)
4. Lowe, D.: Distinctive image features from scale-invariant keypoints. Int. J. Comput. Vision **60**(2), 91–110 (2004)
5. Lobato, F., Jr, V.: Fish swarm optimization algorithm applied to engineering system design. IEEE Trans. Antennas Propag. **11**, 143–156 (2014)
6. Kennedy, J., Eberhart, R.C.: Particle swarm optimization. In: Proceedings of ICNN, vol. 4, pp. 1942–1948 (1995)
7. Eberhart, R.C., Shi, Y.: Particle swarm optimization: developments, applications and resources. In: Proceedings of CEC, vol. 1, pp. 81–86 (2001)
8. Karimkashi, S., Kishk, A.: Invasive weed optimization and its features in electromagnetics. IEEE Trans. Antennas Propag. **58**(4), 1269–1278 (2010)

9. Aydin, I., Karakose, M., Akin, E.: A multi-objective artificial immune algorithm for parameter optimization in support vector machine. Appl. Soft Comput. **11**(1), 120–129 (2011)
10. Sheikh, A., Lye, H., Mansor, S., Fauzi, M., Anuar, F.: A content based image retrieval system for marine life images. In: Proceedings of ISCE, pp. 29–33 (2011)
11. Caicedo, J., Gonzalez, F., Romero, E.: Content-based histopathology image retrieval using a kernel-based semantic annotation framework. Biomed. Inf. **156**(44), 519–528 (2011)
12. Akakin, H., Gurcar, M.: Content-based microscopic image retrieval system for multi-image queries. Proc. TITB **16**(4), 758–769 (2012)
13. Yang, C., Li, C., Tiebe, O., Shirahama, K., Grzegorzek, M.: Shape-based classification of environmental microorganisms. In: Proceedings of ICPR, pp. 3374–3379 (2014)
14. Li, C., Shirahama, K., Grzegorzek, M.: Environmental microbiology aided by content-based image analysis. In: Pattern Anal. Appl. (2015)
15. Yang, C., Tiebe, O., Pietsch, P., Feinen, C., Kelter, U., Grzegorzek, M.: Shape-based object retrieval by contour segment matching. In: Proceedings of ICIP, pp. 2202–2206, Aug 2014
16. Li, C., Shirahama, K., Grzegorzek, M.: Environmental microorganism classification using sparse coding and weakly supervised learning. In: Proceedings of EMR@ICMR, pp. 9–14 (2015)
17. Bosch, A., Zisserman, A., Muoz, X.: Scene classification using a hybrid generative/discriminative approach. IEEE Trans. Pattern Anal. Mach. Intell. **30**(4), 712–727 (2008)
18. van de Sande, K., Gevers, T., Snoek, C.: Evaluating colour descriptors for object and scene recognition. IEEE Trans. Pattern Anal. Mach. Intell. **32**(9), 1582–1596 (2010)
19. Broilo, M., Natale, F.G.D.: A stochastic approach to image retrieval using relevance feedback and particle swarm optimizatio. IEEE Trans. Multimedia **12**(4), 267–277 (2010)
20. Jiao, L., Wang, L.: A novel genetic algorithm based on immunity. IEEE Trans. Syst. Man Cybern. Part A Syst. Hum. **30**(5), 552–561 (2000)
21. Perez, F., Koch, C.: Toward colour image segmentation in analog VLSI: algorithm and hardware. Int. J. Comput. Vision **12**(1), 17–42 (1994)
22. Cheng, H., Jiang, X., Sun, A., Wang, J.: Colour image segmentation: advances and prospects. Pattern Recogn. **34**(12), 2259–2281 (2001)
23. Burt, P.J., Adelson, E.H.: The Laplacian pyramid as a compact image code. IEEE Trans. Commun. **31**(4), 532–540 (1983)
24. Crowley, J.L., Stern, R.M.: Fast computation of the difference of low pass transform. IEEE Trans. Pattern Anal. Mach. Intell. **6**(2), 212–222 (1984)
25. Mikolajczyk, K., Schmid, C.: Scale and affine invariant interest point detectors. Int. J. Comput. Vision **60**(1), 63–86 (2004)
26. Hinchey, M.G., Sterritt, R., Rouff, C.: Swarms and swarm intelligence. Computer **40**(4), 111–113 (2007)
27. Kishida, K.: Property of average precision and its generalization: an examination of evaluation indicator for information retrieval experiments (2005)

Enhancing Spatial Data Warehouse Exploitation: A SOLAP Recommendation Approach

Saida Aissi, Mohamed Salah Gouider, Tarek Sboui and Lamjed Ben Said

Abstract This paper presents a recommendation approach that proposes personalized queries to SOLAP users in order to enhance the exploitation of spatial data warehouses. The approach allows implicit extraction of the preferences and needs of SOLAP users using a spatial-semantic similarity measure between queries of different users. The proposal is defined theoretically and validated by experiments.

Keywords Spatial datacube · Semantic similarity · Spatial similarity · Recommendtion · Personalization

1 Introduction

In recent years, the amount of spatial data is becoming increasingly important due to the development of new technologies for acquiring these data such as sensor networks, remote sensing, etc. The management, exploitation and visualization of spatial data are typically enabled by Geographic Information Systems (GIS). On the other hand, Data Warehousing (DW) and OLAP (Online Analytical Processing) systems allow online analysis of large amounts of alphanumeric data. However, OLAP systems are not intended to manage spatial data. SOLAP systems (Spatial OLAP) combine both GIS and OLAP technologies to integrate spatial data in

S. Aissi (✉) · M.S. Gouider · L. Ben Said
SOIE Laboratory, High Institute of Management, Tunis, Tunisia
e-mail: saida.aissi@isg.rnu.tn

M.S. Gouider
e-mail: ms.gouider@isg.rnu.tn

L. Ben Said
e-mail: lamjed.bensaid@isg.rnu.tn

T. Sboui
ESSPCR (UR 11ES15) & CONTOS2, Faculty of Sciences,
Department of Geology, Tunis, Tunisia
e-mail: Tarek.sboui@ulaval.ca

R. Lee (ed.), *Computer and Information Science*,
Studies in Computational Intelligence 656, DOI 10.1007/978-3-319-40171-3_10

multidimensional databases. SDW (Spatial Data Warehouse) and SOLAP systems provide new opportunities for spatio-temporal analysis of geo-localized information [11].

Spatial data warehouses store enormous amount of data which are historised and aggregated according to several levels of granularity. In addition, spatial data warehouses store both thematic and spatial data that have specific characteristics such as topology and direction. As matter of fact, extracting interesting information by exploiting spatial data warehouses could be complex and difficult; Users might ignore what part of the warehouse contains the relevant information and what the next query should be. SOLAP users can exploit spatial data warehouses by launching a sequence of MDX (Multidimensional Expressions) queries. The user facing a large amount of complex data does not know where to find relevant information, and how to use them. On the other hand, recommendation systems aim to help users navigating large amounts of data. Hence, developing a Spatial OLAP (SOLAP) recommendation system would facilitate information retrieval in spatial data warehouses.

In this paper, we propose to enhance spatial data warehouse exploitation by recommending personalized MDX queries to the user taking into account his preferences and analysis needs. The users' needs and preferences regarding the data stored in the SDW (spatial data warehouse) are detected implicitly during the recommendation process using a spatio-semantic similarity measure.

To the best of our knowledge, there is no proposed similarity measure between spatial MDX queries and no developed recommendation approaches in the field of Spatial OLAP systems taking into account the specific characteristics of spatial data.

More precisely, our contribution consists in (1) proposing a measure of the similarity between spatial MDX queries including a semantic similarity measure and a spatial similarity measure, (2) defining a framework for using these measures to search the log of a SOLAP server to find out the set of relevant queries matching the current query, (3) generating recommendations and classifying the recommended queries in order to present first the most relevant ones, and (4) implementing the approach and evaluating its efficiency.

This paper is organized as follows: Sect. 2 presents a review of literature on OLAP and SOLAP personalization approaches. Section 3 presents a motivating example explaining the usefulness of our proposal. Section 4 presents the proposed spatio-semantic similarity measure between developed MDX queries. Section 5 describes the conceptual framework of the suggested recommendation approach. Section 6 exposes the set of experiments conducted to test the effectiveness of the proposal. Finally, Sect. 7 concludes our work.

2 Literature Review

Personalization is a research topic which has been the subject of several studies in the fields of Information Retrieval [1] as well as in the Web Usage Mining [26]. In recent years, several academic studies have been conducted for personalizing OLAP systems and GIS [2, 3].

Personalization aims to provide quick access to relevant information by eliminating all irrelevant information tailored to the needs, behaviors and preferences of users. Personalization could be realized by adapting the system to users' needs and preferences or by recommending personalized queries to enhance the system exploitation. Personalization in OLAP could be realized at four levels: (i) schema level, (ii) visualization level, (iii) interrogation level and (iv) recommendation level.

Schema level personalization adapts the OLAP system by personalizing the conceptual schema according to users' needs. Schema personalization can be realized by adding new hierarchy level [8, 12] or by filtering facts and/or dimensions [13, 15].

Visualization level personalization concerns data visualization of the result of OLAP queries. The system presents personalized visualization of OLAP query's result according to the preferences of the user and taking into account visual constraints [6].

Personalization of OLAP system at the interrogation level allows the user to express his needs when launching his query on the system. This is realized by proposing new preference algebra allowing the user to express his preferences over the MDX queries [5, 9, 17, 22].

The fourth level of OLAP personalization involves the establishment of query recommendation systems. Recommendation is defined as the process that proposes one or more OLAP queries to users according to their preferences and needs in order to facilitate the analysis process and facilitate the exploration of the OLAP system. Jerbi et al. (2010) define recommendation as a process that exploits users' previous queries on the cube and what they did during the previous session in order to recommend the next query to the current user.

In the context of OLAP, recommendation approaches could be classified into two main categories: (i) collaborative recommendation approaches which is based on query log analysis: the system recommends alternatives queries to help users navigating the cube [14, 20] and (ii) individual recommendation approaches based on user profile analysis: the system provides alternatives and anticipated recommended query taking into account the user context.

Personalization in SOLAP systems has been investigated only in the work of Glorio et al. [15, 16] whom propose an adaptation of data warehouse schema by integrating spatial data at the conceptual level. They offer various types of personalization actions: change of the structure of the data warehouse, adding geometric descriptions for multidimensional elements and adding geometric elements in multidimensional model. Layouni et al. [18] applied the approach proposed by Giacometti et al. [14] to spatial data warehouses. The latter approach proposes a recommendation of OLAP queries by matching the current user session with previous sessions [14]. However, the specific characteristics of spatial data are not considered in the work of [18] (e.g., spatial coordinates and spatial relationships).

Hence, personalization is a search topic which was widely explored in the context of OLAP systems by recommending personalized queries or adapting interfaces and/or contents to the user preferences and needs. However, in the context of Spatial OLAP system and despite the complexity of geospatial data stored in SDW, there is no proposed recommendation approach taking into account the specific characteristic of spatial data in order to facilitate spatial data cube exploitation.

In this section we have reviewed the current approaches for personalization in OLAP and SOLAP systems. In the next sections, we detail our proposal of personalized queries recommendation approach in SOLAP systems. Let's present first an example motivating our proposal.

3 Motivating Example

In this section, in order to illustrate the usefulness of our approach, we present in Fig. 1 an example of a spatial data warehouse described by a multidimensional schema allowing the analysis of the crop (production). The schema diagram is presented using the formalism of Malinowski et al. (2004). It allows the analysis of the weight and the amount of the production according to the dimensions *zone*, *time and product* by answering queries such as: "what is the total production of

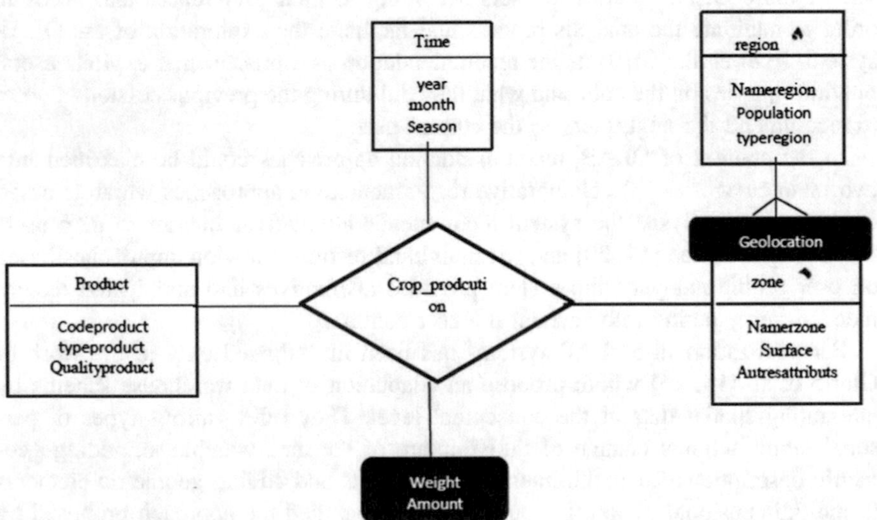

Fig. 1 A cube schema for the analysis of the crop production

biological products in 2014 in the Norths regions?", "what is the total production of high quality products in the suds regions between 2012 and 2014?" The dimension *zone* is a spatial dimension containing a spatial level *region*.

Suppose two users of this spatial data warehouse; a User A and a User B. They are both responsible for agricultural animal production (e.g., red meat, white meat and milk) in the northern regions of Tunisia. The two users have the same preferences regarding the data stored in the data warehouse (i.e., agricultural animal production in the northern region).

The user A launch the following query q_A: "What is the total production of red meat in the area of Beja?". We suppose that the user B has used the system and triggered the following query q_B "what is the total production of milk in the region of Jendouba in 2014?". By analyzing the semantic and spatial similarity between the two user's queries, we observe a semantic link (red meat and milk are two animal-based products) and a spatial link (Beja and Jendouba are two close regions located in the north of Tunisia). Once the user A launches the query q_A, the system we developed recommends the query q_B to this user to assist him/her in the exploration of the spatial warehouse and to accelerate the process of research of relevant information. Thus, by analyzing the semantic and spatial similarity between queries, it is possible to implicitly identify the preferences and the analysis's objectives of the user in order to make useful recommendations.

4 A Spatio-Semantic Similarity Measure Between MDX Queries

The basic idea of the approach is to recommend personalized MDX queries to the current user of SOLAP system. As part of our approach, we propose to detect implicitly users' preferences by comparing the preferences of the current user with the preferences of previous users of the data warehouse. The idea of exploiting the similarity between user's preferences to provide recommendations is a popular technique in collaborative filtering recommendation approaches in several domains such as the classification of opinions, the transactional databases [24] and the traditional non-spatial data warehouses [14, 20].

Queries launched by the user over the SOLAP system are key elements for analyzing users' behavior and preferences. The idea is to identify the similarity between the preferences of SOLAP users throught their MDX queries triggered on the system and to use this similarity to recommend, to the current user, personalized MDX queries. Developing a similarity measure between MDX queries is then a fundamental step in the recommendation process.

4.1 Semantic Similarity Measure for Comparing MDX Queries

To compute the semantic similarity between two spatial MDX queries, we propose to compute the semantic distance between the different references of each query. Several similarity measures between concepts are proposed in the literature. The similarity measures are usually based on knowledge representation model offered by ontologies of concepts and semantic networks [23]. The concepts in our proposal are represented by the query references. To compute the semantic distance between references of each query we use an edge counting method by applying the Rada distance [21] using an application ontology representing the different concepts of the multidimensional database model (dimensions, measures and attributes). The Rada distance computes the minimum number of edges which separate the query references in the ontology. We opted for the Rada distance because it is simple, accurate and efficient [21].

Definition Given $q1$, $q2$: Two spatial MDX queries, where:

$R_1 = \{R_1^1, R_1^2 ... R_1^i, R_1^n, M_1^1, ... M_1^p, ... M_1^m\}$: the set of references of the query q1; $1 \leq i \leq n$ and $1 \leq \mathrm{p} \leq \mathrm{m}$

$R_2 = \{R_2^1, R_2^2, R_2^j ... R_2^n, M_2^1, M_2^k, ... M_2^h\}$: the set of references of the query q2; $1 \leq j \leq n$ and $1 \leq \mathrm{k} \leq \mathrm{h}$.*i, j, p, k, m, n, h are positive integers*

Let $\mathrm{A} = (dr_{ij})_{1 \leq i \leq \mathrm{n},\ 1 \leq \mathrm{j} \leq \mathrm{n}}$: Denote the matrix of the semantic distances between the references $R_1^i (1 \leq i \leq \mathrm{n})$ of the query $q1$ and the references $R_2^j (1 \leq \mathrm{j} \leq \mathrm{n})$ of the queryq2; $1 \leq i \leq \mathrm{n}$ and $1 \leq \mathrm{j} \leq \mathrm{n}$; dr_{ij}: The distance between the reference R_1^i of the query $q1$ and the reference R_2^j of the query $q2$ using the Rada distance, and based on the knowledge representation model offered by the application ontology.

Let $(B = (dm_{pk}))_{1 \leq p\ \leq \mathrm{m},\ 1 \leq \mathrm{k} \leq \mathrm{h}}$: Denote the matrix of the semantic distances between the references M_1^P $(1 \leq \mathrm{p} \leq \mathrm{m})$ of the query q1 and the references M_2^k $(1 \leq \mathrm{k} \leq \mathrm{h})$ of the query q2 (in term of measure); $1 \leq p \leq m$ and $1 \leq \mathrm{k} \leq \mathrm{h}$; dm_{pk}: The distance between the reference M_1^P of the query $q1$ and the reference M_2^k of the query $q2$ using the Rada distance, and based on the knowledge representation model offered by the application ontology.

The semantic distance between the query $q1$ and the query $q2$ denoted *Dsem* ($q1$, $q2$) is obtained as follows:

$$Dsem(q1, q2) = \sum_{i=1}^{n} \sum_{j=1}^{n} \mathrm{drij} + \sum_{p=1}^{m} \sum_{k=1}^{h} dmpk; 1 \leq \mathrm{i,j} \leq \mathrm{n}, 1 \leq \mathrm{p} \leq \mathrm{m\ and}\ 1 \leq \mathrm{k} \leq \mathrm{h} \quad (1)$$

The semantic similarity measure is derived from the semantic distance as follows:

$$Simsem(q1, q2) = 1/(1 + Dsem(q1, q2)) \quad (2)$$

4.2 *Spatial Similarity Measure Between MDX Queries*

Spatial similarity assessment is hard to address because of the complexity of spatial properties and relations. Since it is believed that spatial relations, mainly topology, direction and metric distance, capture the essence of spatial similarity assessment [10, 19]. Hence, the spatial similarity between two MDX queries is defined as follows:

Definition Given $q1$, $q2$: two MDX queries. The spatial distance between $q1$ and $q2$ is computed as follows:

$$D_{Spatial}(q1, q2) = D_{topo}(q1, q2) + D_{dir}(q1, q2) + D_{met}(q1, q2) \quad (3)$$

where:

D_{topo} $(q1, q2)$ the topological distance between $q1$ and $q2$
D_{dir} $(q1, q2)$ the distance in term of orientation between $q1$ and $q2$
D_{met} $(q1, q2)$ the metric distance between $q1$ and $q2$

Details of computing the topological distance, the metric distance and the directional distance between two MDX queries for spatial recommendation is presented and validated in [4].

4.3 *The Spatio-Semantic Similarity Between MDX Queries*

The spatio-semantic distance between two queries represents the degree of spatial similarity and semantic relatedness between them. In assessing similarity between spatial MDX queries we take into account both the spatial and semantic aspects. Thus, the spatio-semantic distance between two queries is derived from the spatial and semantic distance as follows:

Definition Given $q1$ and $q2$ two MDX queries, D_{Sem} $(q1, q2)$ and $D_{Spatial}$ $(q1, q2)$ denote respectively the semantic distance and the spatial distance between the queries $q1$ and $q2$. The spatio-semantic distance and similarity between q1 and q2, denoted resepectively D_{SPS} $(q1, q2)$ *and* S_{SPS} $(q1, q2)$, are computed as follows:

$$D_{SPS}(q1, q2) = \left(D_{Sem}(q1, q2) + D_{Spatial}(q1, q2)\right)/2 \quad (4)$$

$$S_{SPS}(q1, q2) = 1/(1 + D_{SPS}(q1, q2)) \quad (5)$$

5 Personalized Recommendations of SOLAP Queries: Conceptual Framework

The idea of the approach is to recommend personalized MDX queries to the user. The queries are adapted users' preferences and objectives of analysis. Preferences of the user are detected implicitly using collaborative filtering technique by comparing the current user query with previous queries triggered by former users and recorded in a log file. The comparison between the current user query and previous queries is performed using the spatio-semantic similarity measure already presented in the previous section (Sect. 2).

The recommendation approach is based on four main phases: (1) File log filtering, (2) Generation of candidate queries from the filtered log file, (3) Ranking of final recommendations and finally (4) the proposal of relevant queries. Figure 2 explains the theoretical framework of the proposed approach. In this section, we detail the different phases of the recommendation approach we proposed.

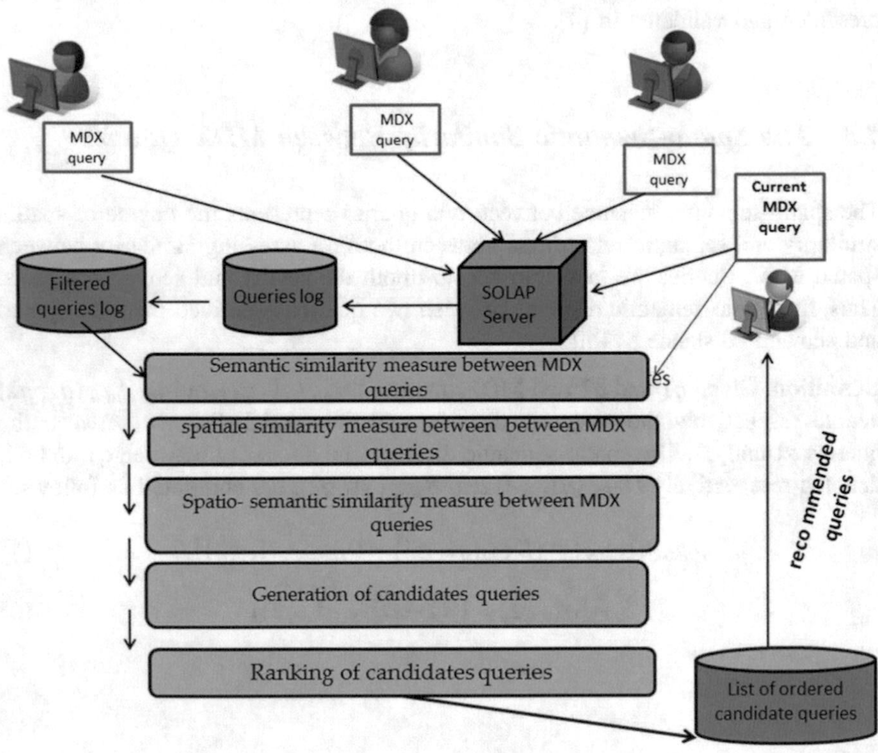

Fig. 2 Theoretical framework of SOLAP recommendation approach

5.1 *Log File Filtering*

The log file containing previous queries already launched on the cube can be very large because of the high number of queries and users. The time of recommendation can significantly increase. To address this problem, we propose to preprocess the log file to remove non-relevant queries in the recommendation process. The filtering criterion of the log file is the execution date (the age) of a query defined as a parameter of this phase to be settled by the user or the administrator of SOLAP system according to his preferences. Only relatively recent queries are considered in the recommendation process.

5.2 *Generation of Candidate Queries*

This phase allows generating all candidates queries for recommendation from the initial log file after preprocessing. Generating candidate queries is based on measuring the spatio-semantic similarity between MDX queries (Sect. 2). The most spatially and semantically similar queries to the current user query are presented in the list of candidate queries. At this level, two methods for generating candidate queries are proposed. The first method is based on the selection of candidate queries having a similarity value, relative to the current query, equal or exceeding a predetermined threshold of spatio-semantic similarity. The spatio-semantic similarity threshold is a parameter defined by the user. The second method is based on the selection of the *k* most similar queries to the current query. The value of *k* is also a parameter specified by the user/administrator according to his preferences.

Each method has its advantages and disadvantages. The first method ensures a good quality of recommendations because queries that do not respect a defined threshold of similarity will be directly eliminated. However, this method may give an empty set of recommendations if the defined value of the threshold similarity is high. As against, the method of the *k* most similar queries allows to guarantee a minimum number of recommendations, however, the quality of a recommendation is not sufficiently controlled.

5.2.1 Candidate Queries Generation Using the Threshold Similarity

This method extracts from the filtered log file the set of candidate queries that are similar to the current query taking into account a predetermined similarity threshold. It ensures a good quality of recommendations because queries that do not respect a defined threshold of similarity will be directly eliminated. However, this method may give an empty set of recommendations if the defined value of the threshold similarity is high. The algorithm CQGS we defined takes as input the filtered log file, the current user query, the SIM function and the similarity threshold s.

SIM function is used to compute the spatio-semantic similarity values between the current query and the queries presented in the filtered log file.

```
Algorithm CQGS (qc, log F, n,SIM, s)
Input
  qc : current user query
  log F: filtered log file
  n : number of queries in the filtered log file
  SIM :a function that computes the spatio-
semantic similarity between two MDX queries
  s :a similarity threshold
Output
  Cquery :The set of candidate queries

  Cquery ← Ø
  For I in 1..n do
  If SIM(qi, qc)≥ s
  then Cquery ← Cquery U {qc}
  end if
  end for
  return Cquery
```

5.2.2 Generation of Top-k Similar Queries

Based on the method of k most similar queries, the number of candidate queries is fixed and the recommendation system generates, from the log file, the k most similar queries to the current user. Regarding this method, no similarity constraint is imposed.

5.3 Ranking of Candidate Queries

Once candidate queries are generated, we classify them to be recommended to the user by order of relevance. At this level, we define three ranking criteria (1) Candidates queries ranking according to their occurrence frequencies in the log file (according to their use frequency), (2) queries ranking according to their execution date (the most recent queries are the most favored), and (3) candidate queries ranking according to the spatial and semantic proximity relative to the current query (Queries having a high similarity are the first to be recommended).

In order to have an efficient recommendation process, we propose a combined ranking method that includes the three ranking criteria. The following ranking order of candidate queries will be applied in the recommendation system: (1) the spatio-semantic similarity relative to the current query, (2) the frequency of use of a query and (3) the execution date of a query. Our choice of this ranking order is motivated by the following reasons:

Spatial and semantic similarity is considered as the most important factor to be used in the recommendation process. Indeed, this criterion reflects the preferences and interests of the user detected through the spatio-semantic similarity measure.

Providing a spatially and semantically relevant query to the current user's one refers to offering a query that responds more to the user needs and analysis objectives.

Also, we have classified the age factor as the last ranking criterion because the original log file have been filtered according to this criterion and only the queries that are relatively recent according to user preferences. Thus, at this level, the use of age factor just makes a final classification of the most relevant candidate queries.

5.4 Proposal of Final Recommendations

After generating and ordering the set of candidate queries, the relevant recommendations should be proposed to the user. At this level, we need to address two points: First, we must specify the maximum number of recommendations to be proposed. Second, we must specify the action to be taken in case the set of candidate queries is empty.

Concerning the first point, we propose to introduce a maximum number of five recommendations for the user. For the following reasons: We believe that when the number of recommendations exceeds five queries, the user will be tempted to read, analyze and compare the proposed queries to make his/her choice. This will cause an increase in the reflection time and affects spatial data cube exploitation. In some cases, the user may not take into consideration the recommendation system when the number of proposals is not relevant.

Regarding the second point, we have considered unnecessary to provide a default recommendation for the user when the set of candidate queries is empty taking into account on the following principle "*It is better not to make a recommendation than to propose an irrelevant recommendation*". This allows keeping a good perception of the recommendation process in the mind of the user and encouraging him/her to take seriously each proposal.

6 Experimental Evaluations

In this section, we present the set of experiments we conducted in order to evaluate the efficiency of our proposal. First, we present the Similarity measure evaluation. Second, we present the prototype RECQUERY that implements the recommendation approach.

6.1 Similarity Measure Evaluation

In order to evaluate the efficiency of the proposed similarity measure, we used the technique of human evaluation based on the Spearman's correlation coefficient. We asked 15 human subjects to assign the degrees of similarity between 30 pairs of spatial MDX queries that have different degrees of spatio-semantic relatedness as assigned by the proposed similarity. Then, we calculate the value of the Spearman's correlation coefficient that express the correlation between the similarity values given for the 30 pairs of queries using our proposal and similarities values provided by human evaluation. The obtained value of the Spearman's coefficient equal to 0.82. This value express that there is a high degree of positive correlation (0.82) between the similarity values accorded to the evaluated queries, using the human evaluation technique and the similarity values accorded to the evaluated queries using our proposal. Hence, the obtained correlation coefficient proves the efficiency of the proposed measure.

We developed the CoSIM (ComputeSIMilarity) system using Java language. CoSIM system implements our proposal of the spatio-semantic similarity measure. It identifies the references of a given MDX query and computes the semantic measure, the spatial measure and the spatio-semantic similarity measure between two MDX queries according to the crop production spatial data warehouse presented in Fig. 1.

6.2 RECQUERY System

In order to evaluate the efficiency of the proposed approach, we developed the prototype RECQUERY (RECommend QUERY). RECQUERY implements all phases of the approach to provide user with useful and relevant queries. In a first phase, the current user runs a query on the spatial datawarehouse stored in the Data Base Management System MySQL, through the GeoMondrian SOLAP server. The previous queries of past users have already been recorded in the log file. In a second phase, the current query of the user and previous queries are loaded into RECQUERY system. During the recommendation process, RECQUERY accesses information in the SOLAP server. Finally, the system recommends to the current user an ordered set of queries.

Hereafter, we present a set of conducted experiments to evaluate the efficiency of the approach. First, the performance of RECQUERY system is tested regarding the time required to make recommendations for different size of the log file. Second, the approach is evaluated using the precision indicator.

6.3 Performance Analysis

For performance analysis, we evaluate the time needed to present recommendations to the user for filtered log files with different sizes. The log size presents the total number of queries contained in the log file after preprocessing. The results of this evaluation are presented in Fig. 3 that shows that the execution time varies linearly

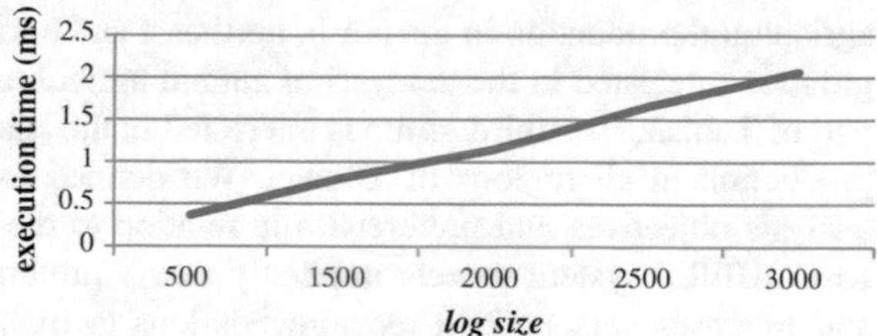

Fig. 3 Performance of RECQUERY

according the size of the original log file. This still acceptable since it reaches a value equal to 2 s for a log file size equal to 3000 queries.

6.4 Precision Evaluation

In this section, we present the set of conducted experiments in order to evaluate the efficiency of our approach using the precision indicator. First, we explain the evaluation technique we applied. Second, we present the experiments and their results.

6.4.1 Evaluation Technique

Precision is an indicator widely used to assess the quality and performance of recommendations in many areas [7]. It reflects the proportion of recommendations that are in the user preferences [7]. In general, the precision is measured as follows:

$$Precision = |\{relevant\ recommendations\} \cap \{proposed\ recommandations\}| \\ /|\{relevant\ recommendations\}|$$

In our case, the proposed recommendations are presented by all the recommended queries (RC_{final}). The relevant recommendations are presented by all recommended queries accepted by the user (RC_{accp}). Thus, the precision is obtained as follows:

$$Précision = \left|\{RC_{final} \cap RC_{accp}\}\right| / \{RC_{final}\}$$

A human evaluation technique is used to calculate precision. 15 human subjects who already manipulate Spatial OLAP systems and the MDX language are formed on the content of the agricultural spatial data warehouse presented in Fig. 1. We distinguished three groups of users. We asked each group to launch a set of queries to achieve some objectives of analysis. Analysis objectives are defined explicitly for each group of users. We asked the first group to focus on the analysis of organic agricultural production in the north, northeast and northwest of Tunisia. The second group is interested in the analysis of animal agricultural production in the southern part of Tunisia. The third group is interested in the analysis of vegetable agricultural production in all regions of Tunisia. We defined explicitly for each group their analysis objectives and preferences in relation to the data in the spatial data cube. RECQUERY system detects implicitly user's preferences and analysis objectives and proposes personalized recommendations to make information retrieval easier and faster by enhancing spatial data cube exploitation. We asked each user to launch 10 queries to exploit the cube and retrieve the needed information.

6.4.2 Precision Evaluation for Different Similarity Thresholds

The aim of this test is to study the efficiency of the recommendation system using the similarity threshold method for the generation of candidate queries. Figure 4 shows the results of this test are presented in Fig. 4 that shows that at least 25 % of the recommendations have been chosen by the user for relevant information retrieval. This precision rate increases proportionally according to the similarity threshold defined by the user. The precision rate reaches a value of 72 % for a threshold similarity equal to 0.8. We also note that the precision varies proportionally to the defined similarity threshold. In fact, the higher is the similarity threshold, the greater are the quality of recommendations and the precision rate. However, for a similarity threshold which exceeds 0.8, the precision value starts to decrease. In fact, when the similarity threshold is relatively high, the number of candidates queries is reduced which implies a reduction of possible choices presented for the user.

We note that the precision value depends also on the quality of the queries already registered in the log file. Indeed, the higher is the number of queries having a high similarity with the current query, the greater is the precision value.

6.4.3 Optimal Number of Final Recommendations

The purpose of this test is to define the optimal number (f) of final recommendations to be presented to the user. To do this, the value of the precision is evaluated according to different numbers of final recommendations. The results of this test are

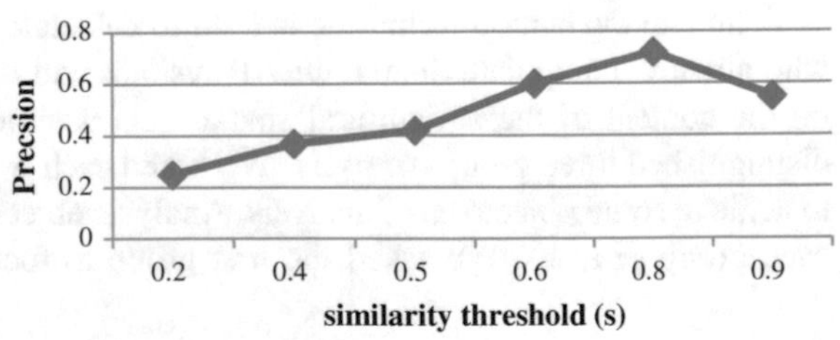

Fig. 4 Precision of RECQUERY for different similarity thresholds

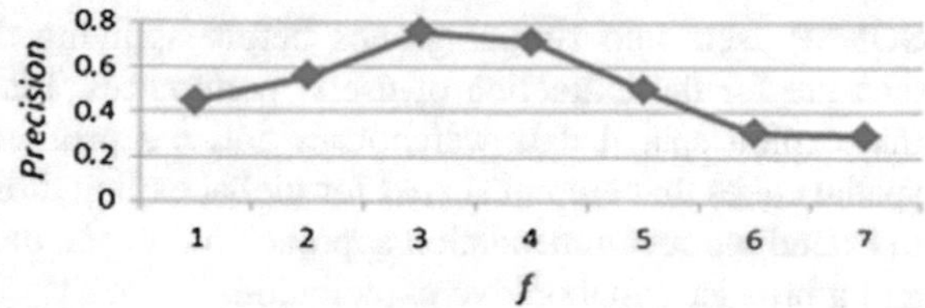

Fig. 5 Optimal number of final recommendation

shown in Fig. 5. The highest value of relevant recommendations is obtained for f = 3. When the number of final recommendations presented to the user exceeds three queries, the precision of the system decreases. In fact, when the user had a large number of recommendations (more than 3 queries), he/she will be disturbed and take more times to analyze and select the proposals. The user may also ignore the recommendation system and build his/her own query. Thus, the optimal number of final recommendations to submit to the user is three queries. We note that the number of final recommendations can slightly vary from one user to another according to his experience with MDX language as well as the structure and the content of the spatial data warehouse.

7 Conclusion

In this paper, we proposed a recommendation approach that assists SOLAP users through the recommendation of personalized MDX queries. The approach detects implicitly users' preferences and analysis objectives using a spatio-semantic similarity measure between MDX queries. We presented a conceptual framework explaining the various phases of the approach as well as the conducted experimental evaluation of the proposal. Experimental results show that efficient recommendations are defined, and in most cases good and helpful recommendations are proposed. In fact, during various tests, at least 25 % of the recommendations have been triggered by the users to advance their search process. This rate is, in most cases, above 40 % and reaches 72 %.

The advantage of the approach is its flexibility since it allows users and administrators to intervene during the different phases of the approach (e.g., to choose the method of candidate queries generation as well as the criteria to be used for candidate queries ranking, etc.), and to fix the value of different parameters (e.g., the similarity threshold and the number of candidate queries to be generated). Hence, the proposed recommendation system could be adapted according to the nature of the application and the data and also according to the user preferences.

As future works, we propose to enhance the implicit extraction of users' preferences by analyzing not just the triggered MDX queries on the system but also the SIG operations launched by the users like pan, zoom and selection on spatial objects. At the recommendation process level, we propose to develop the collaborative filtering recommendation process by performing firstly a clustering of

SOLAP users into similar groups before applying the spatio-semantic similarity measure for the extraction of users' preferences. Finally, organizations and firms that exploit spatial data warehouses notice a proliferation of various versions of spatial cubes that are not shared for global exploitation [25]. It is therefore possible to extend our recommendation approach on several data warehouses for a wider use and a broader collaborative exploitation.

References

1. Adomavicius, G., Tuzhilin, A.: Toward the next generation of recommender systems: a survey of the state-of-the-art and possible extensions. IEEE Trans. Knowl. Data Eng. 734–749 (2005)
2. Aissi, S., Gouider, M.S.: Towards the next generation of data warehouse personalization system: a survey and a comparative study. Int. J. Comput. Sci. Issues **9**(3) (2012)
3. Aissi, S., Gouider, M.S.: A new similarity measure for spatial personalization. Int. J. Database Manag. Syst. **4**(4) (2012)
4. Aissi, S., Gouider, M.S., Sboui, T., Ben Said, L.: Enhancing spatial data cube exploitation: a spatio-semantic similarity perspective. In: ICIST 2014, Communications in Computer and Information Science Series, vol. 0465, pp. 121–133. Springer (2014)
5. Aligon, J., Golfarelli, M., Marcel, P., Rizzi, S., Turricchia, E.: Mining preferences from OLAP query logs for proactive personalization. In: ADBIS, 2011, pp. 84–97 (2011)
6. Bellatreche, L., Giacometti, A., Marcel, P., Mouloudi, H.: A personalization of MDX queries. In: Bases de Données Avancées (BDA'06) (2006)
7. Bellogín, A., Castells, P., Cantador, I.: "Precision-oriented evaluation of recommender systems": an algorithmic comparison. In: RecSys'11 Proceedings of the Fifth ACM Conference on Recommender Systems, pp. 333–33 (2011)
8. Bentayeb, F.: K-means based approach for OLAP dimension updates. In: ICEIS 08, pp. 531–534 (2008)
9. Biondi, P., Golfarelli, M., Rizzi, S.: Preference-based datacube analysis with MYOLAP, ICDE, pp. 1328–1331. Hannover (2011)
10. Bruns, H.T., Egenhover, M.J.: Similarity of spatial scenes. In: Seventh International Symposium on Spatial Data Handling, Delft, The Netherlands, pp. 4A.31–42 (1996)
11. Bédard, Y.: Spatial OLAP. In: 2ème Forum annuel sur la R-D, Géomatique VI: Un monde accessible, Montréal (1997)
12. Favre, C., Bentayeb, F., Boussaid, O.: A user-driven data warehouse evolution approach for concurrent personalized analysis needs. Integr. Comput. Aided Eng. (ICAE) **15**(1), 21–36 (2008)
13. Garrigós, I., Pardillo, J., Mazón, J.N., Trujillo, J.: A conceptual modeling approach for OLAP personalization. In: ER 2009, Conceptual Modeling. LNCS, pp. 401–414, vol. 5829. Springer, Heidelberg (2009)
14. Giacometti, A., Marcel, P., Negre, E., Soulet, A.: Query recommendations for OLAP discovery-driven analysis. In: IJDWM, pp. 1–25 (2011)
15. Glorio, O., Mazón, J., Garrigós, I., Trujillo, J.: Using web-based personalization on spatial data warehouses. In: EDBT/ICDT Workshops (2010)
16. Glorio, O., Mazón, J., Garrigós, I., Trujillo, J.: A personalization process for spatial data warehouse development. In: Decision Support Systems, vol. 52, pp. 884–898 (2012)
17. Golfarelli, M., Rizzi, S.: Expressing OLAP preferences. In: SSDBM, pp. 83–91 (2009)
18. Layouni, O., Akaichi, J.: A novel approach for a collaborative exploration of a spatial data cube. Int. J. Comput. Commun. Eng. **3**(1) (2014)

19. Li, B., Fonseca, F.: TDD—a comprehensive model for qualitative spatial similarity assessment. Spatial Cogn. Comput. **6**(1), 31–62 (2006)
20. Negre, E.: Quand la recommandation rencontre la personnalisation. Ou comment générer des recommandations (requêtes MDX) en adéquation avec les préférences de l'utilisateur. Technique et Science Informatiques **30**(8), 933–952 (2011)
21. Rada, R., Mili, H., Bicknell, E., Blettner, M.: Development and application of a metric on semantic nets. IEEE Trans. Sys. Man and Cybern. **19**(1), 17–30 (1989)
22. Ravat, F., Teste, O.: Personalization and OLAP databases. In: Annals of Information Systems, New Trends in Data Warehousing and Data Analysis, vol. 3, pp. 71–92. Springer (2008)
23. Rezgui, K., Mhiri, H., Ghédira, K.: Theoretical formulas of semantic measure: a survey. J. Emerg. Technol. Web Intell. 5(4), 333–342 (2013)
24. Sapia, C.: PROMISE: predicting query behavior to enable predictive caching strategies for OLAP systems ". In: DaWaK, pp. 224–233 (2000)
25. Sboui, T., Bédard, Y.: MGsP: extending the GsP to support semantic interoperability of geospatial datacubes. In: ER Workshops, pp. 23–32 (2010)
26. Srivastava, J.R., Cooley, R., Deshpande, M., Tan, P.-N.: Web usage mining: discovery and applications of usage patterns from web data. In: SIGKDD Explorations, pp. 12–23 (2000)

On the Prevalence of Function Side Effects in General Purpose Open Source Software Systems

Saleh M. Alnaeli, Amanda Ali Taha and Tyler Timm

Abstract A study that examines the prevalence and distribution of function side effects in general-purpose software systems is presented. The study is conducted on 19 open source systems comprising over 9.8 Million lines of code (MLOC). Each system is analyzed and the number of function side effects is determined. The results show that global variables modification and parameters by reference are the most prevalent side effect types. Thus, conducting accurate program analysis for many adaptive changes processes (e.g., automatic parallelization to improve their parallelizability to better utilize multi-core architectures) becomes very costly or impractical to conduct. Analysis of the historical data over a 7-year period for 10 systems how that there is a relatively large percentage of affected functions over the lifetime of the systems. The trend is flat in general, therefore posing further problems for inter-procedural analysis.

Keywords Function side effects · Pass by reference · Function calls · Static analysis · Software evolution · Open source systems

1 Introduction

It is very challenging to statically analyze or optimize programs that have functions with side effects. Most studies show functions with side effects pose a greater challenge in many software engineering and evolution contexts including system

S.M. Alnaeli (✉) · A.A. Taha · T. Timm
Department of Computer Science, University of Wisconsin-Fox Valley, Menasha, WI 54952, USA
e-mail: saleh.alnaeli@uwc.edu

A.A. Taha
e-mail: RICEA9147@students.uwc.edu

T. Timm
e-mail: TIMMT4191@students.uwc.edu

R. Lee (ed.), *Computer and Information Science*,
Studies in Computational Intelligence 656, DOI 10.1007/978-3-319-40171-3_11

maintainability, comprehension, reverse engineering, source code validation, static analysis, and automatic transformation and parallelization.

For development teams, identifying functions that have side effects is critical knowledge when it comes to optimizing and refactoring software systems due to many reasons including regular adaptive maintenance or for more efficient software.

One example is in the context of automatic parallelization (transformation of sequential code to a parallel one that can efficiently work on multicore architectures), where a for-loop that contains a function call with a side effect is considered un-parallelizable, e.g., cannot be parallelized using Openmp. It has been proved that the most prevalent inhibitor to parallelization is functions called within for-loops that have side effects. That is, this single inhibitor poses the greatest challenge in re-engineering systems to better utilize modern multi-core architectures.

A side effect can be produced by a function call in multiple ways. Basically, any modification of the non-local environment is referred to as side effect [1, 2] (e.g., modification of a global variable or passing arguments by reference). Moreover, a function call in a for-loop or in a call from that function can introduce data dependence that might be hidden [3]. The static analysis of the body of the function increases compilation time; hence this is to be avoided. In automatic parallelization context, despite the fact that parallelizing compilers, such as Intel's [4] and gcc [5], have the ability to analyze loops to determine if they can be safely executed in parallel on multi-core systems, they have many limitations. For example, compilers still cannot determine the thread-safety of a loop containing external function calls because it does not know whether the function call has side effects that would introduce dependences.

As such, it is usually left to the programmer to ensure that no function calls with side effects are used and the loop is parallelized by explicit markup using an API. There are many algorithms proposed for side-effect detection [2, 6], with varying efficiency and complexity.

In general, a function has a side effect due to one or more of the following:

1. *Modifies a global variable*
2. *Modifies a static variable*
3. *Modifies a parameter passed by reference*
4. *Performs I/O*
5. *Calls another function that has side effects*

When it comes to side effect detection, the problem gets worse if indirect calls via function pointers or virtual functions are involved. It is very challenging to statically analyze programs that make function calls using function pointers [7, 8] and virtual methods [9]. A single function pointer or virtual method can alias multiple different functions/methods (some of which may have side effects) and determining which one is actually called can only be done at run time. An imprecise, but still valid, analysis is to resolve all function pointers in the system and then assume that a call using each function pointer/virtual method reaches all possible candidate functions in the program. This, of course, adds more complexity and inaccuracy to static analysis. In general, the problem has been shown to be NP-hard based on the ways function pointers are declared and manipulated in the system [7, 8, 10].

In spite of the fact that a variety of studies have been done using inter-procedural and static analysis on function side effects, to the best of our knowledge, no historical study has been conducted on the evolution of the open-source systems over time in regard to function side effect types distribution on open source systems. We believe that an extensive comprehension of the nature of side effect types distribution is needed for better understanding of the problem and its obstacles that must be considered when static analysis is conducted. We believe that development teams who plan to conduct system restructuring to eliminate function side effects in order to adapt for better exporting of multicore architectures, should know the most prevalent type. This way they can plan properly to get optimal results from their work.

In this study, 19 large-scale C/C++ open source software systems from different domains are examined. For each system, the history of each system was examined based on multiple metrics. The number of functions with side effects are determined for each release. Then each kind of function side effect is determined (pass by reference, modification of global variable, in/out operation, and calling affected function). A count of all types are found and kept.

This data is presented to compare the different systems and uncover trends and general observations. The trend of increasing or decreasing numbers of side effect type is then presented.

We are particularly interested in determining the most prevalent function side effect type that occur in most of general open source applications, and if there are general trends. This work serves as a foundation for understanding the problem requirements in the context of a broad **set** of general purpose applications.

We are specifically interested in addressing the following questions. How many functions and methods in these systems do not have any side effects? Which types of side effects are most frequent? Understanding which side effect occur alone in functions is also relevant. That is very important for many software engineering contexts, e.g., automatic parallelization where functions side effects creates a well-known inhibitor to parallelization [11, 12]. Additionally, we propose and provide some simple techniques that can help avoid and eliminate the function side effects, thus improving overall system maintainability, analyzability, and parallelizability.

This work contributes in several ways. First, it is one of the only large studies on the function side effects distribution and evolution on open source general purpose software systems. Our findings show that modification of global variables and parameters sent by reference represent the vast majority of function side effect types occurring in these systems. This fact will assist researchers in formulating and directing their work to address those problems for better software analysis, optimization, and maintenance in many recent software engineering contexts including source code transformation and parallelization.

The remainder of this paper is organized as follows. Section 2 presents related work on the topic of function side effects. Section 3 describes the functions with side effects and approaches used for the detection and determination, along with all possible limitations in our approach. Section 4 describes the methodology we used in the study along with how we performed the analysis to identify each side effect. Section 5 presents the data collection processes. Section 6 presents the findings of our study of 19 open source general purpose systems, followed by conclusions in Sect. 6.

2 Related Work

There are multiple algorithms used for identifying and detecting function side effects. Our concern in this study is the distribution of side effect types and how software systems evolve over time with respect to function side effect presence, in particular for general purpose large-scale open-source software systems, for better understanding and uncovering any trends or evolutionary patterns. That is, we believe there can be valuable information in determining and predicting the solutions and effort required to better statically analyze those systems written in C/C++ languages and design efficient tools that can help eliminating those side effects for better quality source code.

The bulk of previous research on this topic has focused on the impact of side effects on system maintainability, analyzability, code validation, optimization, and parallelization. Additionally, research continues to focus on improving the efficiency of interprocedural techniques and analyzing the complexity of interprocedural side effect analysis [11, 13, 14, 15, 16]

However, no study has been conducted on the evolution of the open source systems over time with respect to the presence of function side effects and their distribution on source code level as we are going to conduct in this work by examining the history of a subset of systems for each release over a 5-year period.

Cooper et al. [17] conducted a study that shows a new method for solving Banning's alias-free flow-insensitive side-effect analysis problem. The algorithm employs a new data structure, called the binding multi-graph, along with depth-first search to achieve a running time that is linear in the size of the call multi-graph of the program. They proved that their method can be extended to produce fast algorithms for data-flow problems with more complex lattice structures. The study focused on the detection of side effects but did not provide any statistics about the usage and distribution of function side effects on the systems they studied and all software systems in general.

In the context of source code parallelization, most compilers still cannot determine the thread-safety of a loop containing external function calls because it does not know whether the function call has side effects that would introduce dependences. That is, parallelizing compilers, such as Intel's [4] and gcc [5], have the ability to analyze loops to determine if they can be safely executed in parallel on multi-core systems, multi-processor computers, clusters, MPPs, and grids. The main limitation is effectively analyzing the loops when it comes to function side effects especially if function pointers or virtual functions are involved [18].

Alnaeli et al. [11] conducted an empirical study that examines the potential to parallelize large-scale general-purpose software systems. They found that the greatest inhibitor to automated parallelization of for-loops is the presence of function calls with side effects and they empirically proved that this is a common trend. They recommended that more attention needs to be placed on dealing with function-call inhibitors, caused by function side effects, if a large amount of parallelization is to occur in general purpose software systems so they can take better advantage of modern multicore hardware.

However, they have not provided the results that show the distribution of function side effect types in general purpose software systems. The work presented here differs from previous work on open source general purpose systems in that we conduct an empirical study of actual side effects in the source code level and all the potential challenges in this process. We empirically examine a reasonable number of systems, 19, to determine what is the most prevalent function side effect present in open source systems and how open source systems evolve over time with respect to function with side effects.

3 Functions Side Effects

We now describe the way we determine and detect side effects and the approach we followed in dealing with indirect calls that are conducted via function pointers and virtual methods within all detected functions. In general, any execution or interaction with the outside world that may make the system run into unexpected status is considered a side effect. For example, any input/output operation conducted within a called function, or modification of the non-local environment is referred to as side effect [1, 2] (e.g., modification of a global variable or passing arguments by reference).

In this study, a function is considered to have a side effect if it contains one or more of the following: (modification of a global variable, modification of a parameter passed by reference, I/O operation, or calling another function that has side effects.

3.1 Determining Side Effects

To determine if a function/method has a side effect we do static analysis of the code within the function/method. We basically used the same approach in our previous studies, however in this study no user defined function/method is excluded [11, 12]. Any variables that are directly modified via an assignment statement (e.g., x = x + y) are detected by finding the l-value (left hand side variables) of an expression that contains an assignment operator, i.e., =, +=, etc. For each l-value variable it is determined if it has a local, non-static declaration, or is a parameter that is passed by value. If there are any l-value variables that do not pass this test, then the function is labeled as having a side effect. That is, the function is modifying either a global, static, or reference parameter. This type of side effect can be determined with 100 % accuracy since the analysis is done local to the function only.

Of course, pointer aliasing can make detecting side effects on reference parameters and global variables quite complex. Our approach detects all direct pointer aliases to reference parameters and globals such as a pointer being assigned to a variable's address (int *ptr; ptr = &x;). If any alias is an l-value we consider this to cause a side effect. However, we currently do not support full type resolution and will miss some pointer variables. Also, there are many complicated pointer aliasing situations that are extremely difficult to address [18] even with very time consuming analysis approaches. For example, the flow-sensitive and context-sensitive analysis algorithms can produce precise results but their complexity, at least $O(n^3)$, makes them impractical for large systems [18]. As such, our approach to detection of side effects is not completely accurate in the presences of pointer aliasing. However, this type of limited static pointer analysis has shown [19] to produce very good results on large open source systems.

The detection of I/O operations is accomplished by identifying any calls to standard library functions (e.g., printf, fopen). A list of known I/O calls from the standard libraries of C and C++ was created. Our tool checks for any occurrence of these and if a function contains one it is labeled as having a side effect. Also, standard (library) functions can be labeled as side effect free or not. As such, a list of safe and unsafe functions is kept and our tool checks against this list to further identify side effects.

Our detection approach identifies all function/method calls within a caller function/method. The functions directly called are located and statically analyzed for possible side effects through the chain of calls. This is done for any functions in the call graph originating from the calls in the caller function/method. This call graph is then used to propagate any side effect detected among all callers of the function.

Even with our analysis there could still be some functions that appear to have side effects when none actually exist or that the side effect would not be a problem for parallelization. These cases typically require knowledge of the context and problem being addressed and may require human judgment (i.e., may not be automatically determinable). However, our approach does not miss detecting any potential side effects. As such we may over count side effects but not under count them.

3.2 *Dealing with Function Pointers and Virtual Methods*

Our approach for calls using function pointers and virtual methods is to assume that all carry side effects. It is the same approach we used in past studied but this time with more involved functions and methods regardless their locations in the system (my papers). At the onset, this may appear to be a problematic, however conservative, limitation. It is very challenging to statically analyze programs that make function calls using function pointers [7, 8] and virtual methods [9]. A single function pointer or virtual method can alias multiple different functions/methods and determining which one is actually called can only be done at run time. An imprecise, but still valid, analysis is to resolve all function pointers in the system and then assume that a call using each function pointer/virtual method reaches all possible candidate functions in the program. This, of course, adds more complexity and inaccuracy to static analysis. In general, the problem has been shown to be NP-hard based on the ways function pointers are declared and manipulated in the system [7, 8, 10].

Function pointers can come in various forms: global and local function pointers. Global forms are further categorized into defined, class members, array of function pointers, and formal parameters [8]. Our tool, *SideEffectDetector*, detects all of these types of function pointers whenever they are present in a function/method. Pointers to member functions declared inside C++ classes are detected as well. Classes that contain at least one function pointer and instances derived from them are detected. Locally declared function pointers (as long as they are not class members, in structures, formal parameters, or an array of function pointers) that are defined in blocks or within function bodies are considered as simple or typically resolved pointers.

Detecting calls to virtual methods is a fairly simple lookup. We identify all virtual methods in a class and any subsequent overrides in derived classes. We do not perform analysis on virtual methods, instead it is assumed that any call to a virtual has a side effect. Again, this is a conservative assumption and we will label some methods that in actuality do not have a side effect to be a problem. A slightly more accurate approach would be to analyze all variations of a virtual method and if none have side effects then it would be a safe call. However, this would require quite a lot of extra analysis with little overall improvement in accuracy.

4 Methodology for Detecting Function Side Effects

A function or method is considered a pure if it does not contain any side effect. We used a tool, ParaSta, developed by one of the main authors and used in [11, 12], to analyze functions and determine if they contain any side effect as defined in previous section. First, we collected all files with C/C++ source-code extensions (i.e., c, cc, cpp, cxx, h, and hpp). Then we used the srcML (www.srcML.org) toolkit [20,

21, 22] to parse and analyze each file. The srcML format wraps the statements and structures of the source-code syntax with XML elements, allowing tools, such as SideEffectDetector, to use XML APIs to locate such things as function/method implementation and to analyze expressions. Once in the srcML format, SideEffectDetector iteratively found each function/method and then analyzed the expressions in the function/method to find the different side effects. A count of each side effect per function was recorded. It also recorded the number of pure functions found. The final output is a report of the number of pure (clear) functions and functions with one or more types of side effects. All functions were deeply analyzed and side effect types distributions were determined as well.

Findings are discussed later in this paper along with limitations of our approach.

5 Data Collection

Software tools were used, which automatically analyze functions and determines if they contain any side effects. The srcML toolkit produces an XML representation of the parse tree for the C/C++ systems we examined. SideEffectDetector, which was developed in C#, analyze the srcML produced using XML tools to search the parse tree information using system.xml from the .NET framework. The body of each function is then extracted and examined for each type of side effects in functions. For the function, if no side effect exists in a function it is counted as a pure function otherwise the existence of each side effect is recorded. The systems that were chosen in this study were carefully selected to represent a variety of general purpose open source systems developed in C/C++. These are well-known large scale systems to research communities.

6 Findings, Results, and Discussion

We now study the distribution of 19 general purpose open-source software projects. Table 1, presents the list of systems examined along with number of files, functions, and LOCs for each of them.

These systems were chosen because they represent a variety of applications including compilers, desktop applications, libraries, a web server, and a version control system. They represent a set of general-purpose open-source applications that are widely used. We have a strong feeling gained from their popularity in literature that they represent a good reflection of the types of systems that would undergo reengineering or migration to better take advantage of available technologies and architectures, and targeted for regular maintenance, parallelization, evolution, and software engineering processes in general.

Table 1 The 19 open source systems used in the study

System	Language	KLOC	Functions	Files
gcc.3.3.2	C/C++	1,300,000	35,566	10,274
TAO	C++	1,543,805	39,720	10,000
openDDS3.8	C++	326,471	9,185	1,779
ofono	C	242,153	7,331	527
GEOS	C++	173,742	5,738	815
CIAO	C++	191,535	5,937	1,044
DanCE	C++	102,568	5,292	345
xmlBlaster	C/C++	92,929	2,590	429
Cryto++	C++	70,365	3,183	274
GMT	C++	261,121	4,204	290
GWY	C++	392,130	8,340	594
ICU	C++	825,709	15,771	1,719
KOFFICE	C/C++	1,185,000	40,195	5,884
LLVM	C/C++	736,000	27,922	1,796
QUANTLIB	C++	449,000	12,338	3,398
PYTHON	C	695,000	12,824	767
OSG	C++	503,000	15,255	1,994
IT++	C++	120,236	4,220	394
SAGAGIS	C++	616,102	11,422	1,853
TOTAL	–	**9,826,102**	**89,180**	**18,215**

6.1 Design of the Study

This study focuses on three aspects regarding side effect type distribution and evolution in general purpose systems. First, the percentage of functions containing one or more side effects. Second, we examine which side effect type are most prevalent.

Third, we seek to understand when side effects are the sole cause in function affection. That is, function can have multiple side effects and therefore would require a large amount of effort to remove all the side effects. Thus, we are interested in understanding how often only one type of side effect occurs in a functions. These types of functions would hopefully be easier to refactor into something that is pure and simple to analyze. Lastly, we examine how the presence of side effects change over the lifetime of a software system. We propose the following research questions as a more formal definition of the study:

RQ1: What is a typical percentage of function that are pure and clear from any side effects (have no side effects)?
RQ2: Which types of side effects are the most prevalent?
RQ3: What are their distributions? Exclusively and inclusively (affected by only one type of side effects)

RQ4: Over the history of a system, is the presence of function side effects increasing or decreasing?

We now examine our findings within the context of these research questions.

6.2 *Percentage of Pure Functions (has no Side Effects)*

Table 2 presents the results collected for the nineteen open source systems. We give the total number of functions that have side effects along with the percentage of all side effects we detected. Figure 1 shows the percentage of affected functions computed over the total number of function. As can be seen from Fig. 1, affected functions account for between 29 and 95 % of all function and methods in these systems, with an overall average of 62 %. However, in general, the percentage is high for most of the systems. That is, on average, a big portion of the detected functions in these open source systems could hold side effects. This addresses RQ1.

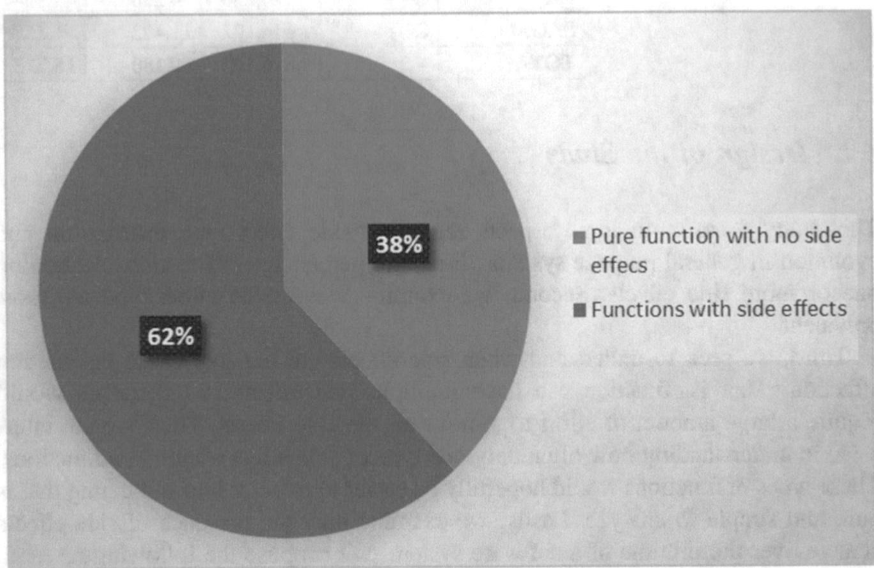

Fig. 1 Total average of pure functions versus affected functions in all 19 systems

6.3 Distribution of Function Side Effects

We now use our finding to address RQ2. Figure 2 presents the details of our findings on the distribution of side effects in studied open source systems. It presents the counts of each side effect that occur within functions. Many of the functions have multiple side effects (e.g., a pass by reference and modification of global variables). As can be seen, modification of global variables is by far the most prevalent across all systems.

For most of the systems this is followed by passing parameters by reference and then calling a function that has a side effect, thus addressing R2. The findings show that DanCE has the lowest percentage which make it a model system when it comes to function side effects addressing through development processes. In contrast, Python seems to have big challenges when it comes to function side effects.

Figures 2 and 3, give the percentage of functions that contain only one type of side effects for each category (addressing RQ3). The average percentage is also given and this indicates that functions that have modification of global variable are clearly more prevalent. We see that CIAO has the largest percentage of modification of global variables as a side effect 25 %, followed by IT++ at 22 %. GWY has the lowest, at 3 %. The percentage of the for-loops that contain only a I/O operation across all the systems is quite small by comparison. Interesting fact here is that passing parameters by reference is considerably high as well for multiple systems as shown in Fig. 3, and Table 2.

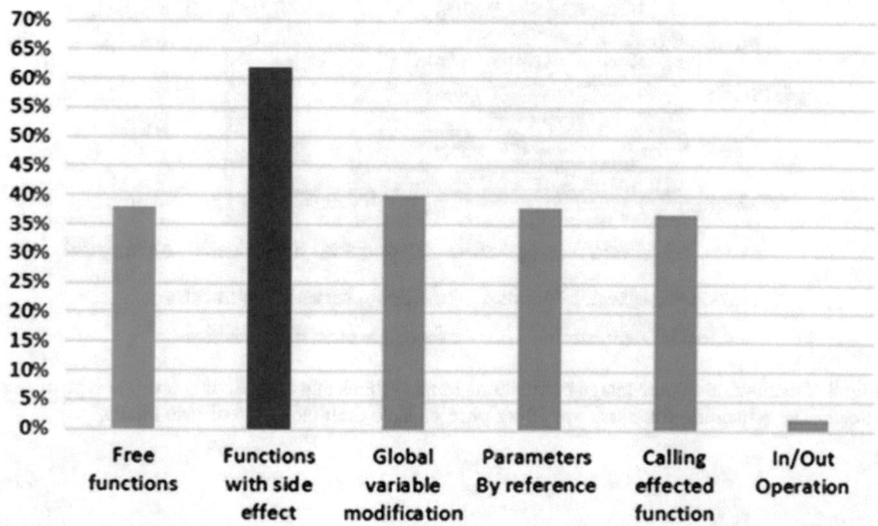

Fig. 2 Average Percentage of functions in all systems that contain only a single type of side effects. The remaining functions are either pure or have multiple types of side effects

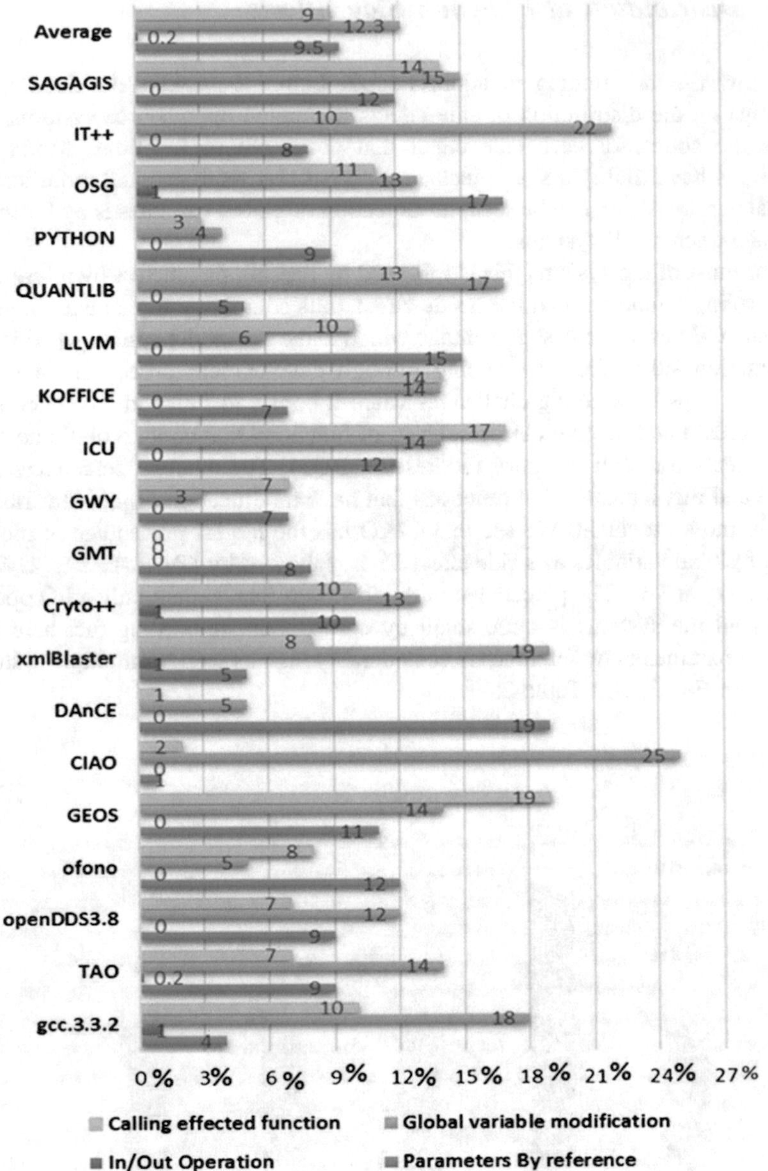

Fig. 3 Distribution of percentage of functions in all systems that contain only a single type of side effects. The remaining functions are either pure or have multiple types of side effects

Table 2 Side effects distribution the 19 open source systems used in the study

System	Functions with side effects	Parameters by reference	In/Out operation	Global variable modification	Calling effected function
gcc.3.3.2	18,803 (52 %)	7,098 (19 %)	1,728 (4 %)	12,685 (35 %)	11,015 (30 %)
TAO	20,349 (51 %)	11,451 (28 %)	361 (<1 %)	13,315 (33 %)	8,135 (20 %)
openDDS3.8	4,414 (48 %)	2,398 (26 %)	244 (2 %)	2,815 (30 %)	2,026 (22 %)
ofono	6,822 (93 %)	5,816 (79 %)	61 (<10 %)	5,283 (72 %)	5,256 (71 %)
GEOS	3,638 (63 %)	1,684 (29 %)	63 (1 %)	1,840 (32 %)	2,515 (43 %)
CIAO	2,187 (36 %)	526 (8 %)	18 (<1 %)	1,953 (32 %)	405 (6 %)
DanCE	1,584 (29 %)	1,217 (22 %)	4 (<1 %)	459 (8 %)	165 (3 %)
xmlBlaster	1,415 (54 %)	630 (24 %)	131 (5 %)	988 (38 %)	810 (31 %)
Cryto++	1,700 (53 %)	834 (26 %)	96	950 (29 %)	1,084 (34 %)
GMT	3,961 (94 %)	3,886 (92 %)	86 (2 %)	3,559 (86 %)	1,919 (59 %)
GWY	7,524 (90 %)	6,619 (79 %)	12 (<1 %)	6,261 (75 %)	5,056 (30 %)
ICU	11,545 (73 %)	6,388 (40 %)	302 (1 %)	6,801 (43 %)	8,572 (54 %)
KOFFICE	21,259 (52 %)	9,109 (22 %)	205 (<1 %)	11,918 (29 %)	13,682 (34 %)
LLVM	15,135 (54 %)	10,080 (36 %)	76 (<1 %)	7,804 (27 %)	10,289 (36 %)
QUANTLIB	6,242 (50 %)	2,344 (18 %)	96 (<1 %)	3,772 (30 %)	3,427 (27 %)
PYTHON	12,282 (95 %)	11,081 (86 %)	183 (1 %)	10,466 (81 %)	10,156 (79 %)
OSG	10,160 (66 %)	6,035 (39 %)	376 (2 %)	5,464 (35 %)	6,643 (43 %)
IT++	2,562 (60 %)	1010 (23 %)	174 (4 %)	1,712 (40 %)	1,646 (39 %)
SAGAGIS	7,089 (62 %)	3,634 (31 %)	65 (<1 %)	3,948 (34 %)	4,715 (41 %)
Average	**62 %**	**38 %**	**2 %**	**40 %**	**37 %**

6.4 Historical Change of Function Side Effect Frequency

Now we present a historical study we conducted on a subset of ten systems chosen from the nineteen studied systems, for 7-year period. Each of the systems, have been under development for 7 years or more. To address RQ4, we examined the versions from 2005 to 2011 (7-year period those 10 systems). Our goal is to uncover how each system evolves in the context of function purity and complexity. Here we measure this by examining the change of side effects within function/methods. Our feeling is that this information could lead to recommendations for utilizing and adapting to the current software and hardware trends.

The change in the percentage of affected function with side effects, and presences of each side effect was computed for each version in the same manner as we described in the previous sections.

These values were aggregated for each year so the systems could be compared on a yearly basis. The systems were updated to the last revision for each year. As before, all files with source code extensions (i.e., c, cc, cpp, cxx, h, and hpp) were examined and their functions were then extracted. Figure 4, presents the change in the percentage of affected functions for each of the 10 systems.

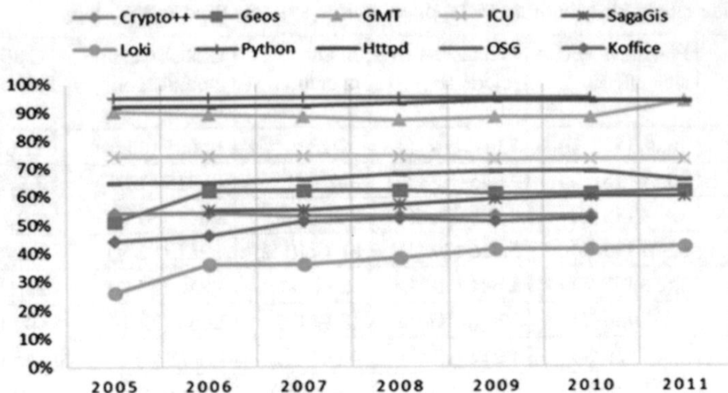

Fig. 4 The evolution of the percentage of functions that has side effects over a 7-year period for the ten systems

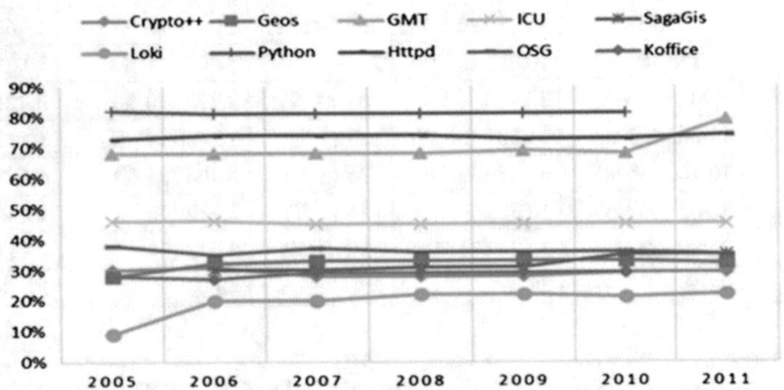

Fig. 5 The percentage of function that modify global variable over a 7-year period for the ten systems

During the 7-year period all systems show a fairly flat trend during the duration. Two systems, OSG and Crypto++, have a steep decline at the end of the period. Geos, Loki, and Koffice have increase for about one year early on and then are relatively flat in proceeding years. Figure 5, presents the percentage of functions that contain a modification of global variable as a side effect. It is approximately a same trend of Fig. 4.

Figure 6, presents the change in the percentage of affected functions by passing arguments by reference for each of the 10 systems. During the 7-year period all systems show a fairly flat trend during the duration except for GMT which starts to increase by mid of 2010 to reach about 90 %. Loki has a steep decline early on and then are relatively flat in proceeding years.

Figure 7, presents the percentage of functions that contain a side effect caused by calling another function that has side effect. The figure shows that a flat trend for

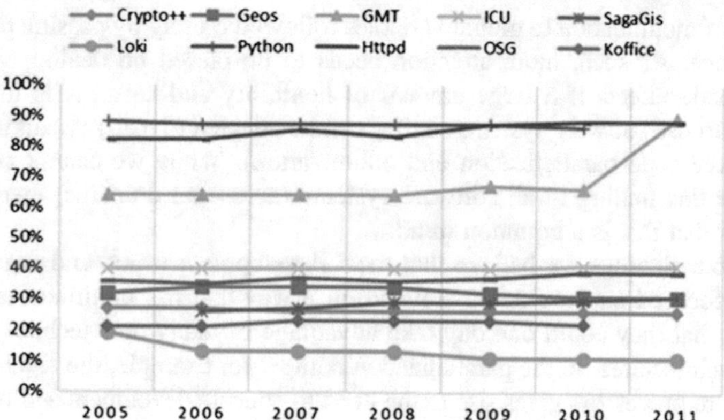

Fig. 6 The percentage of function that had parameters by reference over a 7-year period for the ten systems

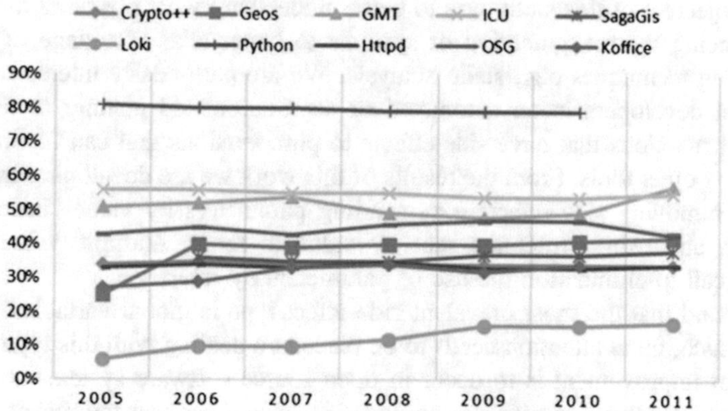

Fig. 7 The percentage of function that calls another affected function over a 7-year period for the ten systems

most of the systems except for Loki which has an identical trend with other side effects. That is, it is approximately a same trend of Figs. 4, 5 and 6.

7 Conclusion

This study empirically examined the distribution and purity of functions (most of affected software engineering contexts by side effects) of nineteen open source general purpose software systems from different system domains. There are no other recent studies of this type currently in the literature targeting function side effects in particular. We found that the greatest side effect of functions is the presence of modification to global variables followed closely by passing parameters by reference. As such, more attention needs to be placed on dealing with those types of side effects if a large amount of flexibility and easiness is to occur in general purpose software systems so they can be adapted to many recent techniques (e.g., source code parallelization and optimization). While we cannot completely generalize this finding to all software systems (across all domains) there is some indication that this is a common trend.

From our findings we believe that most development teams and organizations have not focused on developing software in a way that has minimum use of side effects so that they could one day take advantage of many new technologies e.g., parallel architectures. In the parallelization context for example, the recent ubiquity of multicore processors gives rise to the need to educate developers and make them more aware of the problems that can greatly affect their source code. As we have shown in many studies [11], coding style can play a big role in advancing a system's parallelizability. The software engineering community needs to develop standards and idioms that help developers to avoid the side effects.

The objective of this study was to better understand what obstacles are in place for advancing the reengineering of systems to better take advantage of software engineering techniques e.g., static analysis. We are particularly interested in tools that assist developers in an automated or semi-automated manner to refactor or transform functions that have side effects to pure versions that can facilitate static analysis by other tools. From the results of this work we are developing methods to assist in removing side effects e.g., sending parameters by values instead of by reference, and using structures sent by value to return multiple values from a function call to elimination the use of parameters by reference.

We found that the most prevalent side effect type is global variables modification. As such, more attention needs to be placed on dealing with this type if a large amount of improvement is to occur in open source software systems so they can take better advantage of software engineering techniques and the recent technologies in the market. Our results show some indication that this is a common trend.

Additionally, we empirically showed that coding style can play a big role in advancing a system's maintainability, transformability, parallelizability, and analyzability. That is, developers cause challenges by using parameters by reference in their functions and having their functions modify global variables that can be easily handled outside of the functions. That is at least to some extent due to development teams and organizations not focusing on developing software in a way that could be easily analyzed, comprehended, and maintained. However, the recent challenges in software systems in general when it comes to adaptive changes give rise to the need to educate software developers and engineers and make them aware of the problems that may be caused by using side effects when developing functions.

Acknowledgments This work was supported in part by a grant from the professional development program at University of Wisconsin-Fox Valley and UW-Colleges.

References

1. Ghezzi, C., Jazayeri, M.: Programming Language Concepts. Wiley (1982)
2. Spuler, D.A., Sajeev, A.S.M.: Compiler detection of function call side effects. Technical Report 94/01 (1994)
3. Oracle.: Subprogram call in a loop, http://docs.oracle.com/cd/E19205-01/819-5262/aeuje/index.html (2010)
4. Intel.: Automatic parallelization with intel compilers, http://software.intel.com/en-us/articles/automatic-parallelization-with-intel-compilers/ (2010)
5. Feng, L.: Automatic parallelization in Graphite, http://gcc.gnu.org/wiki/Graphite/Parallelization (2009)
6. Banning, J.P.: An efficient way to find the side effects of procedure calls and the aliases of variables. In: Proceedings of the 6th ACM SIGACT-SIGPLAN symposium on Principles of programming languages, pp. 29–41. ACM, San Antonio, Texas (1979)
7. Cheng, B.-C., Hwu, W.: An empirical study of function pointers using SPEC Benchmarks. In: Proceedings of the 12th International Workshop on Languages and Compilers for Parallel Computing, pp. 490–493. Springer (2000)
8. Shah Anand, R.B.G.: Function pointers in c—an empirical study. Technical Report LCSR-TR- **244**, 11 (1995)
9. Bacon, D.F., Sweeney, P.F.: Fast static analysis of C ++ virtual function calls. SIGPLAN Not. **31**(10), 324–341 (1996)
10. Zhang, S., Ryder, B.G.: Complexity of single level function pointer aliasing analysis. Rutgers University, Department of Computer Science, Laboratory for Computer Science Research (1994)
11. Alnaeli, S.M., Maletic, J.I., Collard, M.: An empirical examination of the prevalence of inhibitors to the parallelizability of open source software systems. Empirical Software Engineering, 1–30 (2015)
12. Alnaeli, S.M., Alali, A., Maletic, J.I.: Empirically examining the parallelizability of open source software system. In: Proceedings of the 2012 19th Working Conference on Reverse Engineering, pp. 377–386. IEEE Computer Society (2012)
13. Chen, K., Lin, J.-Y., Weng, S.-C. Khoo, S.-C.: Designing aspects for side-effect localization. In: Proceedings of the 2009 ACM SIGPLAN Workshop on Partial Evaluation and Program Manipulation, pp. 189–198. ACM, Savannah, GA, USA (2009)
14. Huang, W., Milanova, A.: ReImInfer: method purity inference for Java. In: Proceedings of the ACM SIGSOFT 20th International Symposium on the Foundations of Software Engineering, pp. 1–4. ACM, Cary, North Carolina (2012)
15. Richardson, S., Ganapathi, M.: Interprocedural analysis useless for code optimization. Stanford University (1987)
16. Xu, H., Pickett, C.J.F., Verbrugge, C.: Dynamic purity analysis for java programs. In: Proceedings of the 7th ACM SIGPLAN-SIGSOFT workshop on Program analysis for software tools and engineering. ACM, San Diego, California, USA, pp. 75–82 (2007)
17. Cooper, K.D., Kennedy, K.: Interprocedural side-effect analysis in linear time. SIGPLAN Not. **23**(7), 57–66 (1988)
18. Mock, M., Atkinson, D.C., Chambers, C., Eggers, S.J.: Program slicing with dynamic points-to sets. IEEE Trans. Softw. Eng. **31**(8), 657–678 (2005)
19. Alomari, H.W., Collard, M.L., Maletic, J.I., Alhindawi, N., Meqdadi, O.: srcSlice: very efficient and scalable forward static slicing. J. Soft. Evol. Process (2014)
20. Collard, M.L., Decker, M.J., Maletic, J.I.: Lightweight transformation and fact extraction with the srcML toolkit. In: Proceedings of the 2011 IEEE 11th International Working Conference on Source Code Analysis and Manipulation, pp. 173–184. IEEE Computer Society (2011)

21. Collard, M.L., Kagdi, H.H., Maletic, J.I.: An XML-based lightweight C++ fact extractor. In: 11th IEEE International Workshop on Program Comprehension (2003)
22. Collard, M.L., Maletic, J.I., Marcus, A.: Supporting document and data views of source code. In: Proceedings of ACM Symposium on Document Engineering, p. 8 (2002)

aIME: A New Input Method Based on Chinese Characters Algebra

Antoine Bossard

Abstract Chinese characters are the cement of Asia; they are found in numerous scripts and languages. Yet, the number of characters involved is huge, thus causing a memorisation issue. Both foreign learners and native speakers have to cope with this issue. Aiming at mitigating this issue, we have recently started to describe a novel way to approach Chinese characters: the algebraic way. Such a scientific approach to these characters is innovative in itself, and we propose in this paper a concrete implementation of an input method editor (IME) based on this algebra: *aIME*. Furthermore, we shall experimentally measure the relevance of our IME and its performance by comparing it to several other existing IMEs. From the results obtained, it is clear that the proposed input method brings significant improvement over conventional approaches.

Keywords Language · Script · Combination · Glyph · Interface · Alphabet

1 Introduction

Chinese characters are numerous. It is thus no wonder that their memorisation is challenging, even to native speakers. Character memorisation specifically aimed at foreign learners has been largely discussed in the literature, with various phonetic and non-phonetic methods proposed [1–5]. Character input on a computer is strongly related to the knowledge of such characters, either phonetic knowledge, that is knowing a character's pronunciation, or morphological knowledge, that is knowing a character's shape. Facilitating Chinese characters computer input is thus a meaningful and important issue, which we address in this paper.

A. Bossard (✉)
Graduate School of Science, Kanagawa University,
2946 Tsuchiya, Hiratsuka, Kanagawa 259-1293, Japan
e-mail: abossard@kanagawa-u.ac.jp

R. Lee (ed.), *Computer and Information Science*,
Studies in Computational Intelligence 656, DOI 10.1007/978-3-319-40171-3_12

Major Chinese and Japanese IMEs are phonetic input methods: they work by looking up the pronunciation information input by the user with a database relating characters and their possible pronunciations. Microsoft Pinyin IME for Chinese [6] and Microsoft IME for Japanese [7] are two such method examples. One substantial issue with phonetic IMEs is that the user is required to know the pronunciation of a character in order to input it. And this is a big issue especially for foreign learners: in numerous occasions, they know (or look at) the shape of a character, but are not able to input it easily so as to, for example, look it up in their dictionary for obtaining further information. Now, there exist however other types of IMEs, non-phonetic this time, and thus closer to the one we propose here. Such non-phonetic IMEs are based on character decomposition into basic or key elements (sometimes called primitives).

For instance, the popular Cangjie IME [8] is one of these input methods. This method is complex enough as it requires the user to first remember the basic elements used (i.e. elements that are directly accessible on the keyboard), and also requires the user to know accurately the shape of a character. Because the basic elements used are not the standard radicals, complexity is indeed increased. And because only 24 radicals are used, it is difficult to associate the "forgotten" shapes to these 24 radicals intuitively, thus inducing a heavy learning load onto the user before becoming fluent with the Cangjie IME.

Another non-phonetic famous IME is Wubi [9], also known as Wang Ma [10]. Unlike Cangjie, Wubi input follows the writing order of strokes for a character. One obvious difficulty with Wubi is the keyboard layout complexity. One keyboard key is associated to many different elements, sometimes more than ten (!), thus imposing once again a heavy learning load onto the user.

Another input method is the drawing approach, in use a few modern electronic dictionaries amongst others. This method consists in manually drawing on a touch screen the character, with strokes and order information being use for character look-up. Some electronic dictionaries also enable character look-up by making the user input included basic shapes. These approach are very limited: it is very inefficient (and rather painful) to have to actually draw a character to input it. And shape identification is very imprecise as numerous characters include the same shapes.

Our objective is thus to propose a new, non-phonetic IME to enable input without knowing the pronunciation of a character, while retaining this input method easy enough, for instance by avoiding keyboard clutter as in conventional non-phonetic approaches.

The rest of this paper is organised as follows. We recall in Sect. 2 the character decomposition operations used in this paper. Then, we describe in Sect. 3.1 the overall structure of the proposed input method, and then present in Sect. 3.2 a refined approach. The proposed input method is then empirically evaluated in Sect. 4. Obtained results are discussed in Sect. 5. Finally, this paper is concluded in Sect. 6.

2 Preliminaries

As the proposed system is relying on the previously introduced algebra for Chinese characters [5], we recall in this section important definitions of this Chinese characters algebra.

Considering the set $\mathbb{J}$ of the Chinese characters used in the Japanese language, we have defined the following two composition operations.

Definition 1 ([5]) The operation + realises the horizontal combination of the left operand with the right operand.

$$\begin{array}{rcccl} + & : & \mathbb{J}\times\mathbb{J} & \rightarrow & \mathbb{J} \\ & & \boxed{\text{a}} + \boxed{\text{b}} & \mapsto & \boxed{\text{a b}} \end{array}$$

For example, considering the three characters 木,南,楠 $\in \mathbb{J}$ ("tree", "south" and "camphor tree", respectively), we have 楠 = 木 + 南.

Definition 2 ([5]) The operation × realises the vertical combination of the left operand on top of the right operand.

$$\begin{array}{rcccl} \times & : & \mathbb{J}\times\mathbb{J} & \rightarrow & \mathbb{J} \\ & & \boxed{\text{a}} \times \boxed{\text{b}} & \mapsto & \left|\begin{array}{c}\text{a}\\ \text{b}\end{array}\right| \end{array}$$

For example, considering the three characters 山,石,岩 $\in \mathbb{J}$ ("mountain", "stone" and "rock", respectively), we have 岩 = 山 × 石.

One should note that these two character composition operations + and × have the same evaluation priority, and cover a vast majority of Chinese characters.

3 Methodology

We first describe in this section aIME, the new input method proposed in this paper. We then discuss the various merits of mixing IMEs.

3.1 AIME: A New Input Method

We start by giving details regarding the structure of the proposed input method.

3.1.1 Character Look-Up

The proposed input method is currently relying on two character composition operations: horizontal combination, denoted by +, and vertical combination, denoted by ×. These two character composition operations have been recalled in Sect. 2. For input convenience in aIME, we are using the + and * characters for horizontal and vertical composition operations, respectively. In practice, one character composition expression is input by the user in aIME, for example "木+市". Then, pressing the space bar triggers the look-up of this expression, replacing the input expression by its look-up result, here "柿". We now detail the look-up process.

Character look-up from an algebraic character composition expression is realised by a large database gathering character decomposition information. Each of the two composition operations + and × has its own database. This database is built using nested lists of functional programming (here Scheme), and is structured as shown in Listing 1 (excerpt of the database for the + operation and the one for the × operation). Each database entry is a list whose format is `(left . ((right . result) ... ))` (`left` and `right` to be replaced by `top` and `bottom` for the × database).

Listing 1 Illustrating the character look-up database structure.

```
(define db-plus
 '(
  (木 . ((奇 . 椅) (黄 . 横) (戒 . 械) ... ))
  (口 . ((因 . 咽) (貝 . 唄) (奐 . 喚) ... ))
  ...
 ))

(define db-times
 '(
  (山 . ((風 . 嵐) (厓 . 崖) (屵 . 岸) ... ))
  (宀 . ((女 . 安) (于 . 宇) (豕 . 家) ... ))
  ...
 ))
```

The database for the + operation includes 1,016 "input-able" characters (1,554 characters in total), and the database for the × operation includes 454 input-able characters (732 characters in total). Hence, in total, aIME is relying on a manually assembled database of 1,469 input-able characters, which is made possible by a total of 2,285 characters present in the database.

Expression parsing is conducted as detailed in appendix (and in [11]). Note that here, in aIME, the user need not input parentheses though; the input expression is automatically normalised for later processing by the Racket [12] program statement `(read (open-input-string (~a (string->list (send this get-value))) ))`. Additionally, upon a failed look-up with respect to the database for either the + or × operation, aIME tries as last resort to look up the expression against another database, the radicals database. The purpose of this database is 1) to enable input of character radicals which are not directly accessible on the keyboard (out of 214 radicals, only $47 \times 4 = 188$ are mapped to a keyboard key—we excluded radicals that are rarely used or easily obtained by character composition as explained next), and 2) to speed up input according to user preference: it may be faster to input for instance 丿 × 木 rather than finding 禾 on the keyboard for direct input.

3.1.2 Keyboard Layout

In order to retain high usability, our method uses as basic elements the standard 214 character radicals (部首) as used in all classic dictionaries. The use of standard radicals is very relieving for the user as learners or native speakers are familiar enough with character radicals, and thus it is easy to locate a character on the keyboard since we organise it by stroke number as detailed below. We use the ANSI standard keyboard layout. Each keyboard key is divided into four parts, each accessible with a distinct key modifier as detailed in Fig. 1.

Importantly, the character radicals are arranged in their natural order (as in any classic dictionary) starting from the first keyboard row (QWERTY…), then going on to the second (ASDFG…), third (ZXCVB…) and fourth ('1234…') rows. On each key, considering modifiers, the radicals are ordered as follows: *None*, *Shift*, *Ctrl* and *Alt*. So, for example, the key [Q] enables the input of the very first four radicals 一, 丨, 丶 and 丿, with the modifiers *None*, *Shift*, *Ctrl* and *Alt*, respectively. The final keyboard layout is illustrated in Fig. 2, which also gives an overview of the whole system interface.

It is important to note that several characters that are directly accessible on the keyboard have variants. For instance, the character 水 (access key: [K] has the variants 氵, 氺 and 冫. So, when inputting the character 水, automatic look-up for its variants is performed, which means that the user does not have to pay attention to which variant should be used to form a character.

3.2 *Mixing IMEs*

It actually appears from early experiments that the usability of aIME is maximised when enabling character input not only with the specially laid out keyboard but also

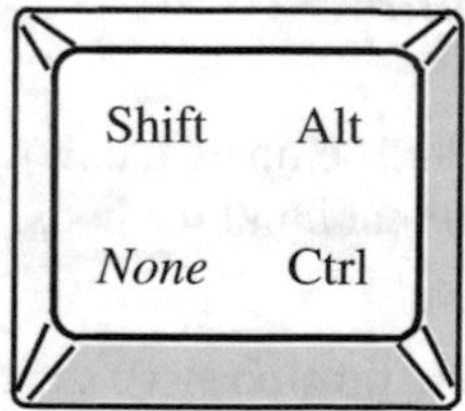

Fig. 1 One keyboard key is assigned four characters. Each of the four characters is accessible with a different key modifier (or none for the *bottom left* character)

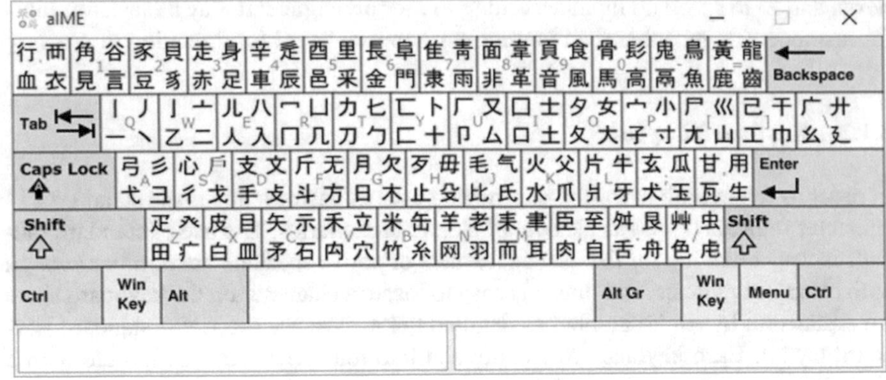

Fig. 2 System interface, including a detailed view of the keyboard layout

with a conventional input method. This way, characters that cannot be directly input with aIME because for instance not suitable for the decomposition operations + and ×, now become accessible.

For example, let us consider the character 榎 ("hackberry tree"). This character can be trivially decomposed as 木 + 夏 = 榎, yet it cannot be directly input with aIME since the character 夏 is neither directly accessible on our keyboard layout nor accessible via further character decomposition.

Here, by using cooperatively aIME and a conventional input method such as the Microsoft IME for Japanese, input of this character 榎 becomes straightforward: by character composition with the + peration, we input the character 木 with aIME (directly keyboard accessible), and the character 夏 with a conventional IME, which will be successfully looked up to 榎 as expected.

This feature is available in our implementation of aIME as shown in Fig. 3, and will be further investigated and quantitatively evaluated in Sect. 4 below.

4 Empirical Evaluation

4.1 Experimental Environment

In this experiment, we measure and compare the time required to input various character sequences when using different input methods. The following five input methods are used:

- Microsoft IME for Japanese (Latin alphabet): standard input method for Japanese included in the Microsoft Windows operating system. Latin alphabet input used to input character pronunciation information. A list of several looked-up characters is then displayed for user selection.

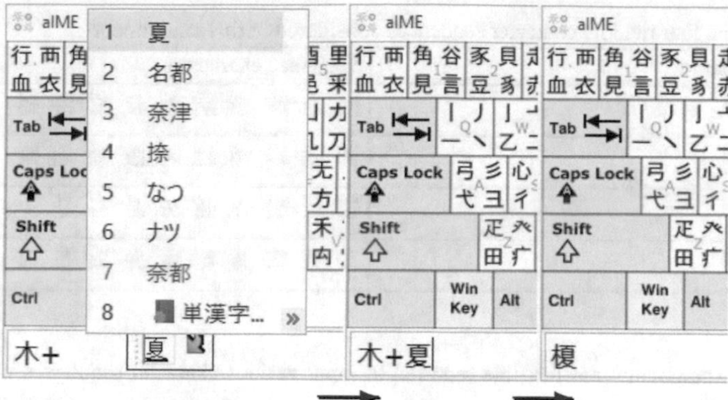

Fig. 3 High usability obtained when combining aIME with a conventional input method. Here, aiming at inputting the character 榎, the character 夏 is input with Microsoft IME for Japanese

- Microsoft IME for Japanese (kana): standard input method for Japanese included in the Microsoft Windows operating system. Direct kana input used to input character pronunciation information. A list of several looked-up characters is then displayed for user selection.
- aIME: exclusive use of the input method proposed in this paper and which is based on algebraic expressions.
- aIME+: cooperation of aIME and a conventional IME as detailed in Sect. 4. Precisely, the conventional input method used was Microsoft IME for Japanese (Latin alphabet); see below.
- Microsoft IME for Japanese (IME Pad): a character is input by manually drawing with a pointing device each stroke included in the character. In most cases, strokes need to be drawn according to the character writing order for a character to be successfully looked up. A list of several looked-up characters is then displayed for user selection.

Because we are measuring input time, and not for example the character set covering rate, trial character sequences were obtained for fair comparison by randomly selecting input-able characters included in the aIME database. The four character sequences used in this experiment are given in Table 1.

The time required to input each of these sequences with each of the five input methods was measured as follows: the clocks is started, and right away the user makes the first input; the clock is stopped when the sequence has been completely processed by the participant.

Lastly, since for fair comparison the user should be experimented enough (and ideally equally) with each of the considered input methods, the author himself will serve as experiment participant.

Table 1 The four random character sequences considered in this experiment

Sequence name	Sequence characters
Sequence 1	伺 詮 鋳 嚇 漸 戠 秋 紅 援 随
Sequence 2	柳 誘 婿 嘲 総 畔 限 効 地 賜
Sequence 3	管 童 墾 昏 蘭 云 易 音 是 少
Sequence 4	条 果 露 秀 努 主 弁 墨 苟 嵐

Table 2 Experimental results: time required to input various character sequences with distinct input methods

Sequence	Input method				
	MS IME (latin)	MS IME (kana)	aIME	aIME+	MS IME Pad
Sequence 1	(1) 2 min 02 s	(1) 2 min 34 s	(4) 2 min 13 s	(2) 1 min 21 s	(0) 2 min 28 s
Sequence 2	(1) 2 min 32 s	(1) 2 min 46 s	(6) 1 min 45 s	(0) 1 min 14 s	(0) 2 min 22 s
Sequence 3	(1) 2 min 09 s	(1) 2 min 23 s	(3) 1 min 10 s	(0) 1 min 33 s	(0) 2 min 14 s
Sequence 4	(1) 2 min 37 s	(1) 2 min 44 s	(2) 1 min 27 s	(0) 1 min 28 s	(0) 2 min 10 s

Numbers in parentheses indicate the number of failed look-ups

4.2 Results

The results obtained from the experiment described in Sect. 4.1 are summarised in Table 2. Columns are labelled as follows: *aIME* for exclusive aIME input; *aIME+* for a cooperation between aIME and Microsoft IME for Japanese (Latin alphabet); *MS IME (Latin)* for Microsoft IME for Japanese (Latin alphabet); *MS IME (kana)* for Microsoft IME for Japanese (kana); *MS IME Pad* for Microsoft IME for Japanese (IME Pad).

We then give in Fig. 4 a summary of look-up measured times in average per character, that is, we only consider successful look-ups. Average time per character per input method as calculated from the values of Fig. 4 is given in Fig. 5.

5 Results Discussion

From the experiment described in Sect. 4, a first result is that character look-up with the proposed aIME input method is instantaneous; there is no delay at all when used on a mid-range machine (Intel i5, 4GB RAM). We now discuss the input time results in details, distinguishing the input method used. We start by giving some general comments, and some comments common to the Microsoft IME for Japanese (Latin alphabet), Microsoft IME for Japanese (kana) and aIME+ input methods.

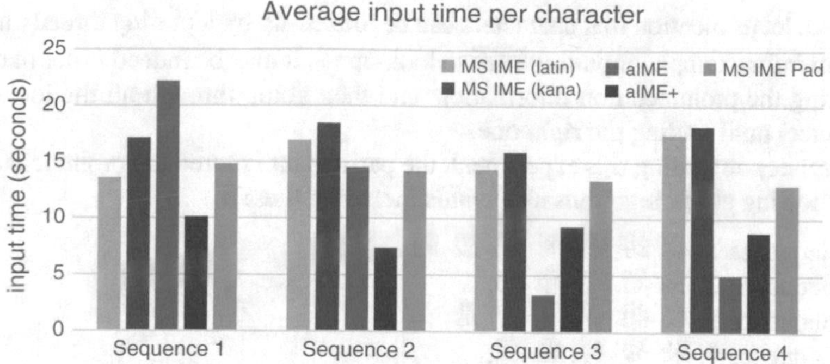

Fig. 4 Experimental results: average time required per character. Only successful look-ups are considered

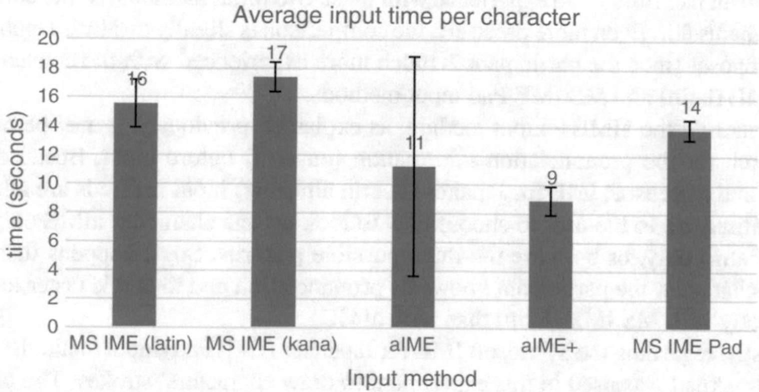

Fig. 5 Experimental results: average time required per character per input method. Standard deviation included

First and foremost, the participant ignores the pronunciation of some of the characters used in this experiment. Hence, the measured time includes the time required to search the pronunciation of such a character. Concretely, the participant uses an online dictionary to find the pronunciation information (onyomi (音読) used in this case in this experiment) for such a character, information which is then used as input to the IME for character look-up. This experiment is thus conducted in a very realistic way as even native speakers ignore the pronunciation of some rather infrequent characters. We recall that aIME we proposed here is a non-phonetic input method, thus allowing to input characters even when ignoring their pronunciation. So, this experiment results are strongly tied to the knowledge of the participant, yet importantly, since the same participant has been involved for all measurements across the considered input methods, this is not an issue: the variation between input times is relevant, while the absolute time values remain linked to the participant profile.

Also, let us mention that characters can be looked-up by inputting directly a sample word: for example, inputting 総務 to look-up 総. It may be indeed faster than first inputting the pronunciation information, and then going through all the looked-up characters until finding the right one.

Precisely, regarding this experiment, the participant ignored the pronunciation of the following characters (thus also words including these):

- in Sequence 1: 詮 鋳 嚇 漸 散 援 随;
- in Sequence 2: 誘 婿 嘲 畔 賜;
- in Sequence 3: 童 墾 昏 易 音 是;
- in Sequence 4: 果 秀 努 墨 苛;

hence having to first find the pronunciation (onyomi) information before input.

Next, regarding Microsoft IME for Japanese (kana) and aIME input methods, the participant had little to no experience with these two input methods, so the comparison remains fair. Even more precisely, the comparison is slightly disadvantageous for our proposal since the participant is much more experienced with the conventional MS IME (latin) and MS IME Pad input methods.

Regarding the aIME+ input method, as explained previously, it may be needed to search for the pronunciation information (onyomi) before input. Because both aIME and Microsoft IME for Japanese (Latin alphabet) input methods are allowed, it is entirely up to the user to choose how to look-up one character: aIME only, MS IME (latin) only, or both are the three possible patterns. So, it happens that for a given character, the participant knows its pronunciation and thus it is faster to input it directly with MS IME (latin) than with aIME.

Lastly, regarding the Microsoft IME for Japanese (IME Pad) input method, a notebook trackpad was used in this experiment to draw characters' strokes. The participant was well experienced with this <trackpad + MS IME Pad> input combination.

Now, regarding the results themselves, two important aspects must be considered: the measured input time obviously, and the characters that were impossible to input using a specific input method. It is easy to see that as expected, combining the proposed aIME with a conventional input method such as MS IME (latin) in our experiment, gives the best results. Indeed, not only do we have input times that are almost halved when using aIME+ compared with MS IME (kana) (e.g. Sequence 4: 1 min 28 s vs. 2 min 44 s, 46 % speed-up), it was in addition possible during that measured time to input significantly more characters than with MS IME (kana); in total for the four sequences, only two characters were not successfully looked-up when using aIME+ against four when using MS IME (kana).

A similar discussion can be made when comparing aIME+ with MS IME (latin), even though the time gap is reduced (e.g. Sequence 4: 1 min 28 s vs. 2 min 37 s, 44 % speed-up, and one character not looked-up with MS IME (latin)), which is mostly due to the fact that the participant is very experienced with this MS IME (latin) input method, compared to the other input methods. In average per character, thus not considering failed look-ups, MS IME (latin) requires 16 s, MS IME (kana) requires 17 s, aIME requires 11 s, aIME+ requires 9 s and MS IME Pad requires 14 s, thus very positive data which shows the significance and relevance of our proposal.

In addition, one should note that when combining aIME with a conventional IME, i.e. aIME+, it was possible to input almost all characters (95 %, against for instance 38 % for aIME only). And this is the reason why recorded average times per character with aIME for Sequences 3 and 4 are smaller than those for aIME+: the number of characters input with aIME is lower than that with aIME+ (a few failed look-ups with aIME), thus shorter time required overall for aIME. Note that because aIME+ includes the aIME input method, it shall by design never took more time than aIME as the user of aIME+ is free to choose to use only aIME input without combining it with MS IME (latin).

Regarding characters that could not be looked-up by aIME, an easy mitigation would be to add support for additional character decomposition operations. For instance, we could add support for 辶-based characters by simply adding definitions in the database with say the + operation: one new entry suffices in the db-plus database, for instance (辶. ((首. 道) ...)) to enable looking up the character 道 with the expression 辶 + 首.

6 Conclusions

Due to their huge number, Chinese characters are very challenging to remember. A same character can have several different pronunciations, and one pronunciation can designate multiple different characters. We have proposed in this paper a new non-phonetic IME for inputting Chinese characters. One important advantage inherent to non-phonetic IMEs is that it is not required any more to know the pronunciation of a character to input it, unlike traditional (phonetic) IMEs such as Microsoft IME for Japanese input and Pinyin IME for Chinese input. Where conventional IMEs, including non-phonetic ones, simply focus on character geometric description, we have proposed in this paper a new IME deeply relying on character relations as described by our character algebra. This has enabled a more accessible IME, with a keyboard layout significantly easier to use and thus a much reduced learning load on the user compared to conventional approaches. In addition, our experiments have shown that the proposed input method aIME when combined with a conventional input method gives the best results, beating previous approaches by almost 40 % in average for the look-up speed of a character.

Future works include gathering metrics regarding the most input keyboard keys and accordingly reorganising the keyboard layout to allow direct access without any key modifier to the most used keys. Also, supporting additional algebraic operations for character decomposition is a meaningful objective, thus aiming at reducing the number of failed look-ups.

Acknowledgments The author sincerely thanks the reviewers for their comments and suggestions.

Appendix—Expression Parsing and Character Look-Up Function

We give in Listing 2 an excerpt of the parsing function used to look up an expression against the database. We are still using the Racket language [12]. The two operations + and × are being handled similarly. This `lookup` function outputs one single character (i.e. the look-up result).

Listing 2 Excerpt of the expression parsing and character look-up function

```
(define (lookup exp)
 (if (= (length exp) 3)
  (let ([c1 (car exp)] [op (cadr exp)] [c2 (caddr exp)])
   (cond [(eq? op '+) ; + operation
          (let ([r1 (assq c1 db-plus)])
           (if r1
            (let ([r2 (assq c2 (cdr r1))])
             (if r2
              (string-ref (symbol->string (cdr r2)) 0)
              ; r2 character look-up failed, try radical look-up
              (let ([r1 (assq c1 db-key-plus)])
               (if r1
                (let ([r2 (assq c2 (cdr r1))])
                 (if r2
                  (string-ref (symbol->string (cdr r2)) 0)
                  #\G)) ; error codes: G, H, I, J, E, @
                #\H))))
            ; r1 character look-up failed, try radical look-up
            (let ([r1 (assq c1 db-key-plus)])
             (if r1
              (let ([r2 (assq c2 (cdr r1))])
               (if r2
                (string-ref (symbol->string (cdr r2)) 0)
                #\I))
              #\J))))]

         [(eq? op '*) ; x operation
          (let ([r1 (assq c1 db-times)])
           (if r1
            (let ([r2 (assq c2 (cdr r1))])
             (if r2
              (string-ref (symbol->string (cdr r2)) 0)
              ; r2 character look-up failed, try radical look-up
              (let ([r1 (assq c1 db-key-times)])
               (if r1
                (let ([r2 (assq c2 (cdr r1))])
                 (if r2
                  (string-ref (symbol->string (cdr r2)) 0)
                  #\G))
                #\H))))
```

```
            ; r1 character look-up failed, try radical look-up
            (let ([r1 (assq c1 db-key-times)])
             (if r1
              (let ([r2 (assq c2 (cdr r1))])
               (if r2
                (string-ref (symbol->string (cdr r2)) 0)
                #\I))
               #\J))))]

         [else #\E]) ; undefined operation
        )
  #\@)) ; ill-formed expression
```

References

1. Richmond, S.: A re-evaluation of kanji textbooks for learners of Japanese as a second language. J. Fac. Econ. KGU **15**, 43–71 (2005)
2. Vaccari, O., Vaccari, E.E.: Pictorial Chinese-Japanese Characters: A New and Fascinating Method to Learn Ideaographs. C. E. Tuttle Co., Tokyo (1958)
3. Heisig, J.W.: Remembering the Kanji, Volume 1: A complete course on how not to forget the meaning and writing of Japanese characters. University of Hawaii Press, Honolulu (2007)
4. Henshall, K.G.: A Guide to Remembering Japanese Characters. C. E. Tuttle Co., Tokyo (1988)
5. Bossard, A.: Premises of an algebra of Japanese characters. In Proceedings of the Eighth International C* Conference on Computer Science & Software Engineering, pp. 79–87, Yokohama, Japan, July 2015
6. Microsoft: Microsoft Pinyin IME 2010 Official Website (in Chinese). http://www.microsoft.com/china/pinyin/ (2010). Accessed 23 April 2016
7. Microsoft: Microsoft IME 2002 seminar text—Introduction to Japanese input. Nikkei Business Publications Inc, Tokyo (2001) (in Japanese)
8. Chu, B.-F.: 辶十首 Wisdom Journey—Autobiography (3): Hot summer (1973–1995). China Times Publishing Co., Taiwan (1995) (in Chinese)
9. Wicentowski, J.: Wubizixing for speakers of English. Yale University, New Haven. http://www.yale.edu/chinesemac/wubi/xing.html (1996). Accessed 1 Nov 2015
10. 北京王码创新网络技术有限公司: Wang Ma Net (in Chinese). http://www.wangma.com.cn/ (2015). Accessed 1 Nov 2015
11. Bossard, A.: Implementation proposal for automatic processing of the algebra on J. In Proceedings of the First International Conference on Computer Application Technologies, pp. 18–23, Matsue, Japan, September 2015
12. Flatt, M.: Creating languages in Racket. Commun. ACM **55**(1), 48–56 (2012)

Author Index

R. Lee (ed.), *Computer and Information Science*,
Studies in Computational Intelligence 656, DOI 10.1007/978-3-319-40171-3

Printed in the United States
By Bookmasters